# THE DEATH OF BUSINESS AS USUAL

ROBERT BRUCE

Copyright © 2020 Robert Bruce

All rights reserved.

ISBN: 9798580040516

# DEDICATION

To those who want to change the world for the better

# CONTENTS

|   | | |
|---|---|---|
| | Acknowledgments | i |
| | Introduction | Pg 7 |
| 1 | Factors for a Successful and Sustainable Business | Pg 11 |
| 2 | Influencers | Pg 57 |
| 3 | Learning from History | Pg 95 |
| 4 | Measuring Sustainable Success | Pg 129 |
| 5 | Global Warming and the State of Earth | Pg 165 |
| 6 | The Future - What can be done | Pg 201 |
| 7 | The Future - What is likely to happen | Pg 239 |
| | References | Pg 271 |
| | About the Author | Pg 321 |

# ACKNOWLEDGMENTS

I wish to thank the following for their support in getting the book to press.
Sheree for researching and editing
Keeran for editing, cover design and resolving IT issues
Pieter Deon for reviewing and making recommendations.

# INTRODUCTION

In 1972, 'The Limits to Growth' (1book) was published, summarizing the Club of Rome's Project on the Predicament of Mankind. There were three key conclusions:

1. *If the present growth trends... continue unchanged, the limits to growth on the planet will be reached within the next 100 years.* **The most probable result will be a rather sudden and uncontrollable decline in both population and industrial capacity.**

2. *It is possible to alter these growth trends and to establish a condition of ecological and economic stability that is sustainable far into the future. The state of global equilibrium could be designed so that the basic material needs of each person on earth are satisfied and each person has an equal opportunity to realize his individual human potential.*

3. *If the world's people decide to strive for this second outcome rather than the first, the sooner they begin working to attain it, the greater will be the chances of success.*

Now, nearly 50 years later, the Covid-19 pandemic is giving mankind an opportunity to stop the mad exponential growth, which is very likely to end in conclusion 1, and enable us to focus on conclusion 2. This gives us the opportunity to make a step change to balance what we are taking from mother Earth with what we can give back.

It goes without saying that this Corona Virus (Covid-19) has had an unprecedented impact on the world economy. I started writing this book at the beginning of 2019 and briefly touched on viruses as just one of many threats, not merely to companies, but to mankind. With hindsight, the Spanish flu of 1918 and the more recent AIDS, EBOLA, MERS and SARS epidemics should have been recognized as warnings of bigger things to come.

The Covid-19 attack has given us pause. Is the focus on a 'growth' economy going to destroy us? How much longer is such a growth economy sustainable? Given that this model has been crippled by recent events, can we find a new one? What would this look like?

In my view, clearly **'business as usual'** pursuits should be in their death throes as we start the slow recovery from the damage wreaked by this virus. The world is simply not the same as it was pre-Covid-19.

More recently, the quest for short term profits has dominated the global economy. However, some institutional shareholders are now demanding long term sustainability from companies in line with established Environmental Social and Governance (ESG) principles. The 'success' of companies no longer depends on short term profits, but on long term sustainability.

The exponential growth required for increased profit and the resulting increased Gross Domestic Product (GDP) cannot continue unabated. This trend has to be replaced. An alternative approach to ensure the survival of the human species is required and it should be requisite for companies to cater for the basic needs of people. One idea is to replace the focus on GDP with the Social Progress Index (SPI). This index captures outcomes related to the UN's 17 Sustainable Development Goals and is a comprehensive snapshot of a country's overall progress towards the achievement of these goals.

At company level, the replacement of untenable short term profit by long term sustainable profit is essential. This includes embracing a sustainable approach and reducing carbon footprints to lower global warming.

In this book there are examples of major influences on businesses, environmental catastrophes and social damage caused by companies, as well as instances of corruption that have destroyed them. Yet there are also examples of businesses and states that are embracing change for the better to ensure carbon neutrality by 2050. I review the current measurements of country and company performance and discuss alternatives to ensure sustainability. The influence of legislation, groups and individuals determines the extent to which businesses will change, with an urgent need to reduce global warming being the most compelling argument. What circumstances are required to shift from doing **business as usual**? I believe the categories for measuring a successful company over the long term are:

- Corporate governance - the framework in which a company operates
- Sustainability - a long term perspective
- Corporate ethics - the elimination of corruption
- Preparing for the unexpected - business strategic and continuity planning.

These points will form the basis of my discussion.

# THE DEATH OF BUSINESS AS USUAL

# CHAPTER 1: FACTORS FOR A SUCCESSFUL AND SUSTAINABLE BUSINESS

## Introduction

In this chapter I elaborate on what I define as "business as usual" and its defining characteristics. Discussion then centres on what constitutes a successful company and what actions destroy a company's credibility and reputation. I also discuss what I believe are the key factors for success which include:

1. Good governance

2. Sustainable business model

3. Ethical leadership and eradication of corruption

4. Expecting the unexpected.

*Box 1.1* gives clarification for discussion in this chapter.

Box 1.1: Types of companies and their relationship to government regulation

> For our purposes, I've defined a company as *"an enterprise which undertakes certain endeavors, usually with the aim of making a profit"*. All legitimate companies need to be registered within a state or country and can be either public or private. A public company is listed on a stock exchange in a certain country and has to disclose investment information to its shareholders, whereas private companies do not. State Owned Enterprises (SOE's) are responsible to a minister of a government, and need to be audited by external auditors, as do public companies. It's also important to keep the link between countries and companies in mind, since the rules and regulations established by governments to regulate businesses often play a key role in how "good" a company can be. Governments need to set basic regulations to promote a climate for fair competition.

## What is 'Business as Usual'?

**Business as usual** infers continuing on the same course as before by maximizing short term profit, usually by depleting resources that don't affect the company directly and often doing catastrophic damage in the process. These include depleting fossil fuel natural resources, destroying the livelihoods of communities, exploiting employees and undertaking corrupt and monopolistic practices. **Companies are not required to balance what they take from the Earth with what they return.** In the process of maximizing profit, a number of companies resort to bribery and corruption. I believe that these businesses are bound to fail in the long run, but not before they do lasting, if not irreparable, damage.

A number of large monopolies and near monopolies have ridden roughshod over communities and the environment, often using corrupt practices to buy concessions from governments, particularly for finite resources. Oil has been the driving force of the modern economy and companies such as Standard Oil, Shell and British Petroleum had dominated the scene until the 1970's when nationalization of oil resources began to temper their control.

Modern day near monopolies such as Microsoft and Google have not had quite the same devastating effect on communities and the environment as 'big oil'. In recent times, monopolistic practices for competitive advantage are being viewed as unacceptable by many governments.

I outline examples of bad traits of 'business as usual' grouped as follows.

1. Poor governance and corrupt leadership

2. No regard for sustainability by destroying the environment and abusing the community in which the company operates

3. Unethical behaviour involving corruption, loss of reputation, product failure, legal disputes, and human rights violations.

I've tried to arrange examples of bad traits into these groups, but there is some overlap.

## 1. Poor governance and corrupt leadership

Certain oil and gas companies, such as Shell and Petrobras, have built reputations as leaders in corporate governance, only to be let down by members of their Boards of Directors. Clean-ups sometimes ensue, but it takes a long time to rebuild a stellar reputation and regain stakeholder trust.

There has been political interference in the affairs of Petrobras in Brazil together with allegations of corruption.

> 'According to trial testimony, top executives at Petrobras accepted huge bribes from a cartel of companies, enriching themselves while also channeling funds to political figures and to the leftist Workers' Party. Although no testimony has emerged suggesting that Ms. Rousseff personally profited from the scheme, she was the chairwoman of the oil giant from 2003 to 2010, roughly corresponding to the period that the system of collusion, kickbacks and payoffs took shape.' – The New York Times *(12news)*.

Ms. Rousseff, then President of Brazil, was impeached and removed from office and the opposition won the subsequent election.

> *'Petrobras has finally announced a positive annual result, five years after the Lava Jato corruption probe left the Brazilian state-controlled company with spiraling debts.'*

In 2002, Shell's Annual Report overstated its reserves of oil and gas thereby falsely influencing its share price.

> 'Ultimately, the alleged overstatement would prove to be 4.47 billion Barrels of Oil Equivalent, *or about 23% of the company's total. A joint investigation was conducted by the* US Securities and Exchange Commission *and* UK Financial Services Authority, *and Shell settled claims with the regulators for USD 120 million and GBP 17 million (USD 28 million), respectively, without admitting to, or denying the findings of the Commission'*.

The Group Chairman, Chief Financial Officer, and Exploration and Production Chief Executive Officer left the company shortly after the reserves revelations (13news) (14news). In spite of it all, Shell remains one of the top earning companies in the world.

## 2. No regard for sustainability

The types of companies that have perpetrated the worst damage tend to be in the energy or chemical sectors. The most extreme of these include:

1. Exxon
2. British Petroleum (BP)
3. Texaco/ Chevron
4. Du Pont
5. Tokyo Electric Power Company (TEPCO)
6. Shell Nigeria.

I discuss these in more detail in **Chapter 3 "Learning from History"**.

A class action in the UK against Cape plc, a British company mining asbestos in South Africa, resulted in a payout of millions (see **Box 3.3**), and another representative action succeeded against DuPont for the pollution of the Ohio River (see **Chapter 3 Learning from history**).

**Box 1.2** outlines a successful class action against the tobacco industry (31wiki) in the US.

> Box 1.2: The Tobacco Master Agreement
>
> *The Tobacco Master Settlement Agreement (MSA) was entered in November 1998, originally between the four largest US tobacco companies (Philip Morris, R.J.Reynolds, Brown & Williamson and Lorillard) and the attorneys general of 46 states.*
>
> *These states settled their Medicaid lawsuits against the tobacco industry to cover their tobacco-related health-care costs. In exchange,*

the companies agreed to curtail or cease certain tobacco marketing practices, as well as to pay, in perpetuity, various annual payments to the states to compensate for some of the medical costs incurred in the care of people with smoking-related illnesses.

The money also funds a new anti-smoking advocacy group that is responsible for such campaigns as Truth. The settlement also dissolved a number of groups supported by the tobacco industry such as The Tobacco Institute, The Center for Indoor Air Research, and The Council for Tobacco Research. In the MSA, the original participating manufacturers (OPM) agreed to pay a minimum of $206 billion over the first 25 years of the agreement.

### 3. Unethical behaviour and corrupt practices

I cover corruption in more detail in **Chapter 3,** but I'd like to mention a few ethics-related examples here.

#### Loss of reputation

Some companies go into bankruptcy, others get bought out, some change their name and others just weather the storm. If a company has enough money to pull though after a major incident or loss of reputation it may yet flourish. An example is BP which is still flourishing even after causing the biggest oil spill in history, not to mention numerous other environmental and safety disasters. Texaco, which was responsible for the biggest oil related damage to rain forests in Ecuador, was absorbed by Chevron who denied any responsibility for the company that they had taken over. As a rule, a company inherits the debts of a company that it acquires. Surely an environmental disaster of this magnitude should be regarded as an inherited debt? (See **Box 3.1.**)

As a result of losing their reputation after the class action of 2017, Du Pont Teflon Production Unit changed its name to Chemours (15movie). (See **Chapter 3 paragraph "Lessons to be learned from serious incidents that have occurred - water and food resources"**).

Hitachi Power Africa (HPA) changed their name to Mitsubishi Hitachi Power Systems Africa (MHPSA) after the corruption scandal uncovered by the US Securities Exchange Commission (SEC). (See **Chapter 3 Learning from history**).

When Samsung S7 phones started to ignite spontaneously they were quickly pulled from the market, but not before passengers, who had these phones in their possession, had them confiscated at the gate before they could board their flights. Samsung is also under investigation in France for allegedly using child labour (16news)(32news). Nevertheless, Samsung remains one of the top selling cell phone companies worldwide.

In an attempt to keep up with Airbus, Boeing (17audio), the American aerospace manufacturer, upgraded its 737 to the 737 Max. Disastrously, they failed to disclose vital software changes to pilots and after two air calamities, these planes were grounded. The American Federal Aviation Authority (FAA) is the aviation regulator, but is so resource restrained that it relies heavily on Boeing to self-regulate. Problems relating to quality management at the new 787 factory in South Carolina are now being revealed following additional revelations regarding faulty lithium batteries in 787 Dreamliners. Seemingly, the focus in the factory has been on production targets and financial incentives for management with quality and safety running a poor second. A Quality Manager reported a number of issues and was moved aside, but became a whistleblower after retirement. When asked if he would fly in a 787, he categorically stated "no".

Glencore plc is a British-Swiss multinational commodity trading and mining company, which has been in the news for questionable dealings in Colombia, Zambia and the Democratic Republic of Congo (DRC). (The latter being one of the world's poorest countries). The resignation of Glencore's Head of Oil followed the publication of alleged scandals (18wiki)(19news).

There appears to be a mantra in business and political circles that goes

"When your reputation is being tarnished, call in the Public Relations experts". London-based television content producer FactBased Communications (FBC Media)(20news), which recently produced programs to spruce up the image of Malaysian politicians, is now under administration. FBC created at least four promotional videos for Malaysia, including controversial issues such as its contentious palm oil industry, their treatment of indigenous people and deforestation.

Bell Pottinger (21news), a British Public Relations company, collapsed in 2017 as a result of South Africa's main opposition party, the Democratic Alliance, accusing Bell Pottinger of exploiting racial tensions, and reporting the scandal to the British Public Relations Association. Bell Pottinger Asia (22news) has now changed its name to Klareco Communications. Some of their less savoury clients included:

- Oakbay Holdings, owned by the Gupta brothers, in support of ex-President Zuma of South Africa
- President Bashar al-Assad, President of of Syria
- Aleksandr Lukashenko, the current president and dictator of Belarus
- General Augusto Pinochet, the former dictator of Chile
- The Uzbek Government.

Strategic Communication Laboratories/ SCL Elections/ SLC Group/ Cambridge Analytica (23movie) has influenced elections in Italy, Latvia, Ukraine, Albania, Romania, South Africa, Nigeria, Kenya, Mauritius, India, Indonesia, The Philippines, Thailand, Taiwan, Colombia, Antigua, St. Vincent & the Grenadines, St. Kitts & Nevis, and Trinidad & Tobago. The Facebook–Cambridge Analytica data scandal was a major political embarrassment in early 2018 when it was revealed that Cambridge Analytica had harvested the personal data of millions of peoples' Facebook profiles without their consent and used it for political advertising purposes. In May 2018, SCL Group and Cambridge Analytica ceased operations.

## Product failure

The US's three largest drug distributors and two manufacturers have agreed with multiple states on a framework to resolve thousands of opioid addiction cases with a settlement worth nearly $50 billion in cash and addiction treatments (40news)(41audio).

The US Food and Drug Administration (FDA) has been confronted with several issues related to leading pharmaceutical companies (24movie), (often referred to as "Big Pharma"), many of which include approval of medical devices and drug testing prior to marketing. The FDA has come under fire for poor regulation, such as accepting company medical trials on face value due to lack of FDA resources. Here are a few examples to illustrate the point:

Johnson and Johnson has been in hot water with respect to a surgical mesh product as well as their talcum powder. The embedded mesh has allegedly caused rejection symptoms and is impossible to fully remove whilst their talcum powder apparently showed traces of toxic asbestos. Johnson & Johnson paid $4.7bn damages in the talc cancer case (25news) and has been instructed to pay $9.9 million for failing to disclose the risk of its surgical mesh devices (26news).

Bayer promoted a contraceptive device for insertion in the uterus (27news) which has caused complications in many patients and some find difficult to remove.

Hip joint manufacturers have also had product failure issues (28news). Certain patients who had undergone hip surgery and received prosthetic joints, were diagnosed with cobalt poisoning. Apparently, wear on the joint produced a chrome cobalt paste that was absorbed into the blood, affecting the brain and poisoning the body.

Toyota was found to have faulty electronically operated accelerators which caused a number of deaths and resulted in payouts for compensation.

Volkswagen's unethical activities (79movie) relating to emission testing of their diesel car engines was a major set back for the company. The CEO resigned and at least 6 executives were charged in the USA.

### Human rights abuses

The use of child labour and slavery to produce goods gives unethical companies unfair advantage in the market place. Other than government regulation, shareholder and customer pressure is essential to limit the use of children as labourers.

The alleged exploitation of child labour by Samsung (32news) was brought up in the French courts by Sherpa and ActionAid France in June 2018.

## What is a successful and sustainable business?

In my view, a "good" company has a sound governance structure, a sustainable business model, highly ethical leadership and strong morals, both to succeed in its venture and to counter corruption. It should also have strategies in place to manage unexpected events such as natural disasters. I give more details in the following paragraphs.

In order to be sustainable, the short term pursuit of profit only, is no longer valid, but a fair profit over an extended period is justifiable to ensure the growth of a for-profit business. Sustainable and successful companies consider not only their shareholder interests but also the natural and social environment in which they operate, as well as their staff and local communities.

It is difficult to change from being highly profitable fossil fuel companies to sustainable successful companies unless shareholder pressure is brought to bear to reduce their carbon footprint. (Some Liquefied Natural Gas (LNG) companies have been making profits of up to 45% after tax on their operations!) This has happened in a number of cases to European oil companies, including Shell and BP, where shareholders have encouraged them to diversify from their fossil fuel interests while still maintaining good profits. A balance between profit and sustainability is necessary to ensure long term survival of the companies. Many US oil and gas giants have not had this pressure to change and so are continuing their 'business as usual' strategies.

The US Forbes magazine, partnered with Just Capital, produces the 'Just 100' survey, which asks respondents what they want to see from the USA's biggest businesses. These are some of the most popular answers:

1. Pay workers fairly (and look after your staff)
2. Treat customers well and protect their privacy
3. Produce quality products
4. Minimize environmental impact
5. Give back to those communities that the company operates in
6. Commit to ethical and diverse leadership
7. Create abundant job opportunities to ensure sustainability.

I look at some companies that have flourished based on these 'Just 100' survey categories.

## Pay workers fairly (and look after your staff)

I have found a few examples of companies who make their staff a priority.

Glassdoor (2news), a website where employees can anonymously rate companies, provides a list of the best companies to work for. In 2019 the US list included information technology companies Facebook and Google, consultants Bain and Boston, as well as Southwest Airlines. Richard Branson, the founder of Virgin, has a mantra 'look after your staff and they will look after the customers'(3book).

Brazil's Recardo Semler (1tedtalk) (4book) revolutionised the approach to managing a company. After assuming control of Semler & Company

(Semco) from his father in 1980, Semler began a decades-long quest to create an organization that could function without him, by studying and then implementing what could best be called "corporate democracy", allowing employees to design their own jobs, select their supervisors, and define pay levels[4].

Hamdi Ulukaya, founder of yogurt manufacturer Chobani (5news), has created a vision for a new "anti-CEO playbook" that prioritizes people over profits (6tedtalk). Ulukaya took over a yogurt factory being closed by Kraft Foods and turned it into one of the biggest yogurt manufacturers in the world.

## Treat customers well and protect their privacy

Apple, Amazon, Toyota, Southwest Airlines and IKEA are considered great at customer service.

Spar, a supermarket chain in South Africa, has a brilliant returns policy: they guarantee to pay double your money back if you are not satisfied with their house-brand products.

## Produce quality products

To draw from a few Asian examples, Huawei, Samsung and Hyundai have grown their market share exponentially due to their leadership in technology. Huawei has grown to be the leader in 5G phone systems while Samsung is one of the biggest mobile phone manufacturers and is involved in the production of a range of electronic items. Hyundai produced its first car in 1968 and is now the 5th largest vehicle manufacturer in the world. It is also a leader in heavy industry and has produced the biggest Liquefied Natural Gas (LNG) tankers to date: Qflex and Qmax.

## Minimize environmental impact

Kimberly-Clark, the manufacturers of Kleenex tissues, states that it obtains 89% of its fibre for tissue products from environmentally preferred sources. (See **Box 4.5.**)

An Irish company, SSE Airtricity (7website), is promoting 100% renewable energy.

Mainstream Renewable Power (8website) has developed, built and operated wind farms for furniture retailer IKEA in Ireland and Canada.

A list of companies who actively attempt to address climate change is given on the UN climate summit website (see also **Chapter 4 paragraph 2 "Sustainable business model"**).

## Give back to those communities in which the company operates

Microsoft supports NGOs serving communities in their localities. For example, its housing project has assisted with Seattle's housing shortage (9news).

Social responsibility has become a big part of IKEAs business and since it's a private company, the owners are able to set the direction of the company without having to consider the short term wishes of shareholders. The company has invested in a number of social projects through its foundation (10website). It's also conscious of being a responsible customer to the many suppliers it has around the world and tries to ensure that the sources of its products support communities and minimize environmental impact. IKEA Foundation has produced cleverly designed tents for refugee camps (84news).

## Commit to ethical and diverse leadership

Ethical leadership is essential to establish appropriate controls to prevent corruption and the abuse of power. Company board diversity and the separation of the roles of Chairman and CEO are key factors in ensuring ethical leadership. Ethics are discussed in detail later in this chapter.

Create abundant job opportunities to ensure sustainability

A company that wishes to create job opportunities has to ensure growth over the long term. Volkswagen (VW) (11website) takes such a long term approach and its investors understand that theirs is a long term commitment. The German State of Lower Saxony, where VW is based, has shares in VW. The union in VW has representation on the management committee (board) to ensure greater job security for their workers.

The internet has created numerous new businesses. Companies such as Amazon have created jobs in the supply and distribution of goods. Information technology companies have created numerous new IT opportunities, although they have not always shown to be sustainable.

It is my belief that, once the the Covid-19 pandemic has abated, future sustainable businesses will be more attuned to providing basic needs, rather than luxury goods and services.

## Key factors for sustainable success

I have identified four key factors for success and described them in more detail:

1. Good governance
2. Sustainable business model
3. Ethical leadership and eradication of corruption
4. Expecting the unexpected.

### 1 Good governance

Codes

*"Corporate governance involves a set of relationships between a*

*company's management, its Board, its shareholders and other stakeholders. Corporate governance also provides the structure through which the objectives of the company are set, and the means of attaining those objectives and monitoring performance are determined."* (33website)

Good governance codes are underpinned by core principles which were established by The Organization for Economic Co-operation and Development (OECD) (34website). OECD is an international economic organization of 34 countries founded in 1961 to stimulate economic progress and world trade. It is a forum of countries committed to democracy and the market economy, providing a platform to compare policy experiences, seek answers to common problems, identify good practices, and co-ordinate domestic and international policies of its members. The core principles of corporate governance practices - Fairness, Accountability, Responsibility and Transparency - are relevant across a range of jurisdictions. These core principles must guide the interpretation of various governance structures and systems at the corporate level and enable the discipline to adhere to them (see Box 1.3).

### Box 1.3: Core principles explained

*Fairness*

*All decisions taken, processes used, and their implementation will not be allowed to create unfair advantage to any one particular party.*

*Accountability*

*Identifiable groups within the organization, for example, governance boards, who take actions or make decisions, should be officially recognized and held accountable for their actions.*

*Responsibility*

*Each contracted party is required to act responsibly to the organization and its stakeholders.*

*Transparency*

*All actions implemented and their decision support will be available for inspection by authorized organization and provider parties.*

*Independence*

*All processes, decision making, and mechanisms used will be established so as to minimize or avoid potential **conflicts of interest**.*

*Discipline*

*All involved parties will have a commitment to adhere to procedures, processes, and authority structures established by the organization.*

Further to OECD's initiative, Extractive Industries Transparency Initiative (EITI) (35website) produced a global standard to promote open and accountable management of natural resources. It seeks to strengthen government and company systems, inform public debate, and enhance trust. In each implementing country it is supported by a coalition of governments, companies and civil society working together. The 12 Principles of EITI, which were agreed by all stakeholders in 2003, lay out the general aims and commitments by all stakeholders.

Good governance includes key company statements: mission, vision, values and code of conduct. It also requires compliance with Governance Codes of a country. Well known codes include Sarbanes Oxley Act (2002) in the USA, UK Corporate Governance Code (2010) and King (IV) Corporate Governance Code in South Africa (2016). Japan published its Corporate Governance Code in June 2015.

It's fairly true to say that when there are rules, there are also those who skirt around them, which can result in a major failure of ethical standards, as occurred in the World Financial Collapse in 2008. US and UK regulatory systems focus primarily on the shareholder, while others, such as the German, Japanese, and South African systems, focus on a greater balance of interests between shareholders and other stakeholders.

King IV (2016) has a primary emphasis on leadership, sustainability, and corporate citizenship. The code is designed on an "apply and explain" basis. This provides boards with the freedom to apply the recommendations differently, or apply other practices, if they consider them to be in the best interests of the organization, but they are then required to justify any departures from the recommendations.

Corporate governance guru, Bob Garratt, outlines his view of US Corporate Governance in **Box 1.4**. Garratt also proposes a new look at Governance Codes as they currently appear to be ineffective. In his view, the relationship between directors, owners, legislators and regulators needs to be addressed and in his latest book, outlined in **Box 1.5**, he discusses Corporate Codes and why many are not working.

A case in point is shareholder Carl Icahn taking the Occidental Board to court, since the CEO would not consider his board member proposals. He remarked that Occidental lacked effective corporate governance (36news). As a result, the Occidental Board was changed.

### Box 1.4: Corporate Governance basket case (37book)

Extract from 'The Fish Rots from the Head' Bob Garratt

*'The US is teetering on the edge of becoming a corporate governance basket case for three reasons:*

1. It is hamstrung by its constitution which is loaded towards the individual states
2. The overweighting importance of legislation in just one state-Delaware
3. Focus mainly on shareholder proxies.'

### Box 1.5 'Stop the Rot: Reframing Governance for Directors and Politicians' Bob Garratt (38book)

*'Stop the Rot' sets governance in a much wider social context. The acceptance of global Human Values in all of our organizations, with their necessary ethics and behaviours, ensures the development of Inclusive Capitalism to the advantage of all.'*

After the Asian financial crisis of 1997-8 (39website), the Association of South East Asian Countries (ASEAN) established a Corporate Governance Scorecard and Award. **Box 1.6** outlines the crisis, also known as the Asian Flu. The ASEAN Corporate Governance Scorecard is discussed in **Box 1.7**.

### Box 1.6: Asian Flu

*The 1997–98 Asian financial crisis began in Thailand and then quickly spread to neighbouring economies. It began as a currency crisis when Bangkok unpegged the Thai baht from the US dollar, setting off a series of currency devaluations and massive flights of capital. The Indonesian rupiah, the South Korean won and the Malaysian ringgit quickly lost significant value.*

*Collectively, the economies most affected saw a drop in capital inflows of more than $100 billion in the first year of the crisis. Significant in terms of both its magnitude and its scope, the Asian financial crisis became a global crisis. The International Monetary Fund (IMF) stepped in to initiate a $40 billion program to stabilize the currencies of South Korea, Thailand, and Indonesia.*

*Though the crisis is generally characterized as financial / economic, it can also be seen as a crisis of governance at all major levels of politics: national, regional and global. In particular, the Asian financial crisis revealed that governments were not adequately performing their historical regulatory functions and were unable to regulate the forces of globalization or the pressures from international actors.*

### Box 1.7: ASEAN Corporate Governance Scorecard

*The Asian Development Bank (ADB) and the ASEAN Capital Market Forum (ACMF) together developed the ASEAN Corporate Governance Scorecard. This provides a rigorous methodology benchmarked against international best practice - including the OECD's principles of corporate governance - to assess the corporate governance performance of publicly listed companies (PLCs) in the six participating ASEAN member countries.*

*This common methodology provides foreign investors and external fund managers comparable information to form part of their investment decision-making process. The scorecard also provides assurance to foreign investors that corporate governance is a priority in the region.*

## Social responsibility

Social responsibility is an integral part of good governance and includes being mindful of support for suppliers, staff, customers and the community in which the company operates.

Microsoft's social responsibility program (9news) recently included funding aspects of the Seattle housing shortage. Pretoria Portland Cement in South Africa strengthens its social responsibility base (40website) by supporting surrounding rural communities.

Lack of social responsibility was highlighted when, with a prime focus on profitability, Kraft Foods decided to shut down a yogurt factory in 2005 that had been operating in northern New York State for generations, and throwing 55 employees out of work. The plant was revived by Hamdi Ulukaya (5news), a Turkish American, in 2007. His company, Chobani, now has the world's largest yogurt plant and employs 600.

## Shareholder vs. stakeholder focus

Focusing on shareholder interests only can have devastating long term effects on a company.

Shareholder focus in British Petroleum, under the Chairmanship of Sir John Browne, led to some major incidents, one being the Deepwater Horizon Oil Spill (see **Chapter 3**). **Table 1.1** discusses the issues of Shareholder vs. Stakeholder focus, whereas the debate between shareholder and stakeholder **value** is discussed in **Box 1.8**.

## Table 1.1: Shareholder vs. Stakeholder focus

|  | Shareholder focus | Stakeholder focus |
|---|---|---|
| **Benefits** | For Investors:<br>• High Rate of Return (RoR)<br>• Reduced risk | For Investors:<br>• Closer monitoring of management<br>• Longer term decision horizon |
|  | For the Economy:<br>• Encourages entrepreneurship<br>• Encourages inward investment | For Stakeholders:<br>Deterrent to high risk decisions |
|  | For Management:<br>• Independence |  |
| **Disadvantages** | For the Economy:<br>• Risk of short-termism<br>• Top management greed | For the Economy:<br>• Reduced financing opportunities for growth |
|  |  | For Management:<br>• Potential interference<br>• Slower decision making<br>• Reduced independence |

## Box 1.8: Shareholder value and/or stakeholder value? (41book)

Extract

*"...there are far more groups apart from shareholders that appear to hold a legitimate 'stake' in the corporation since their interests are already protected in some way. There are not only legally binding contracts to suppliers, employees, or customers, but also an increasingly dense network of laws and regulations enforced by society...*

*,...if a firm closes a plant in a small community and lays off the workers, it is not only the relation with the employees that is directly affected – shop owners will lose their business, tax payments to fund schools and other public services will also suffer – but since the company has no contractual relation to these groups, the traditional model suggests that these obligations do not exist...*

*,...shareholders often buy shares for speculative reasons, and it is the*

*development of the share price that is their predominant interest – and not 'ownership' in a physical corporation."*

Since the 1980s, many US businesses have focused on making short term profits and, with the motto 'greed is good', corporate raiders, like Carl Icahn, have bought companies, down sized them to make big profits, and then sold them off. Fortunately, a wider stakeholder focus is starting to take off in the USA. US Business Round Table (BRT) (42news) (43news) is a lobbying organization for major businesses which consists of 180 plus CEOs . Their new approach is shown in **Box 1.9.**

## Box 1.9: BRT Statement on the Purpose of a Corporation (extract) (44website)

*"While each of our individual companies serves its own corporate purpose, we share a fundamental commitment to all of our stakeholders. We commit to:*

- *Delivering value to our customers. We will further the tradition of American companies leading the way in meeting or exceeding customer expectations.*
- *Investing in our employees. This starts with compensating them fairly and providing important benefits. It also includes supporting them through training and education that help develop new skills for a rapidly changing world. We foster diversity and inclusion, dignity and respect.*
- *Dealing fairly and ethically with our suppliers. We are dedicated to serving as good partners to the other companies, large and small, that help us meet our missions.*
- *Supporting the communities in which we work. We respect the people in our communities and protect the environment by embracing sustainable practices across our businesses.*
- *Generating long-term value for shareholders, who provide the capital that allows companies to invest, grow and innovate. We are committed to transparency and effective engagement with shareholders.*

*Each of our stakeholders is essential. We commit to deliver value to all of them, for the future success of our companies, our communities and our country."*

## Benefits of a good governance framework

A framework for good governance has significant benefits, the primary ones being:

> 1. Effective decision-making based on a high quality knowledge base including a dynamic risk register and monitoring of top performance indicators
> 2. Single source database eliminating duplication of (possibly conflicting) data resulting in reduced cost of data storage and handling
> 3. Increased security for critical data
> 4. Streamlined **Processes** within clearly defined **Systems**, eliminating separate Processes in functional structures
> 5. Common staff focus on the **Company Vision and Customer Satisfaction** (alignment) and the **Strategies for Improvement** of performance, thus increasing productivity and eliminating departmental silos.

## 2 Sustainable business model

Sustainability is part of good governance, where the interest of the shareholders and other stakeholders are balanced to promote long term sustainability of the business. Most businesses are in for the long haul and a sustainable business model, therefore, is essential.

> *"Sustainability is a process or state that can be maintained at a certain level for as long as is wanted."*

The basic principles of sustainability are:

- To do no harm through our activities
- To replace what we have taken
- Where it is not possible to replace, then substitute something of

equal value to our community and the physical environment.

In 1983 the Brundtland Commission was created by the United Nations to find ways in which to save the human environment and natural resources, and prevent deterioration of economic and social development.

> Development that "meets the needs of the present without compromising the ability of future generations to meet their own needs." Brundtland Commission definition of Sustainable Development (45website).

In 1995 the World Business Council for Sustainable Development (WBCSD) (46website) was established. WBCSD is a global, CEO-led organization of over 200 leading businesses working together to accelerate the transition to a sustainable world. WBCSD targets the realization of the **Sustainable Development Goals (SDGs)** through six work programs to achieve systems transformation. I've listed them in **Box 1.10**.

> Box 1.10: Sustainable Development Goals Work Programs
>
> **1. Redefining value** helps companies measure and manage risk, gain competitive advantage and seize new opportunities by understanding environmental, social and governance (ESG) information.
> **2. People program** provides solutions that support companies in ensuring that they remain in tune with the needs, rights, goals and aspirations of society.
> **3. Food, Land & Water program** develops solutions to address key challenges of food & land use systems: food and nutrition security, smallholder livelihoods, natural resource efficiency, including water management, climate change impact and adaptation - using comprehensive approaches and new technologies.
> **4. Climate & Energy program** facilitates interaction on cutting-edge climate and energy topics between WBCSD members, their peers and stakeholders as they address critical industry issues and share best practices and solutions.

**5. Cities and Mobility program** is in the areas of housing, building efficiency, mobility, water and sanitation.

**6. Circular economy program**, aims to bring circularity into the heart of business leadership and practice.

In July 2000, the United Nations (UN) Global Compact (47website) was launched. It is a UN initiative to encourage businesses worldwide to adopt **sustainable** and social responsibility policies, and to report on their implementation. The UN Global Compact is a principle-based framework for businesses, stating 10 principles in the areas of human rights, labor, the environment, and anti corruption. Under the Global Compact, companies are brought together with UN agencies, labor groups, and civil society. The 10 principles are listed in ***Box 1.11***.

## Box 1.11: UN Global Compact Principles

### Human Rights

*Businesses should:*
*Principle 1: support and respect the protection of internationally proclaimed human rights;*

*Principle 2: make sure that they are not complicit in human rights abuses.*

### Labor Standards

*Businesses should uphold:*
*Principle 3: the freedom of association and the effective recognition of the right to collective bargaining;*

*Principle 4: the elimination of all forms of forced and compulsory labor;*

*Principle 5: the effective abolition of child labor; and*

*Principle 6: the elimination of discrimination in employment and occupation.*

### Environment

*Businesses should:*
*Principle 7: support a precautionary approach to environmental challenges;*

*Principle 8: undertake initiatives to promote environmental responsibility; and*

*Principle 9: encourage the development and diffusion of environmentally friendly technologies.*

### Anti-corruption

*Businesses should:*
*Principle 10: Businesses should work against corruption in all its forms, including extortion and bribery.*

In 2006 the heads of leading institutions from 16 countries, representing more than $2 trillion in assets owned, officially signed the Principles for Responsible Investment. It consists of an international network of investors working together to put the six principles into practice. These principles are listed in **Box 1.12**.

*"These Principles for Responsible Investment grew out of the understanding that while finance fuels the global economy, investment decision-making does not sufficiently reflect environmental, social and corporate governance considerations -— or put another way, the tenets of sustainable development"* (48website)

> ### Box 1.12: Principles for Responsible Investment
>
> *Principle 1: We will incorporate ESG issues into investment analysis and decision-making processes.*
>
> *Principle 2: We will be active owners and incorporate ESG issues into our ownership policies and practices.*
>
> *Principle 3: We will seek appropriate disclosure on ESG issues by the entities in which we invest.*
>
> *Principle 4: We will promote acceptance and implementation of the Principles within the investment industry.*
>
> *Principle 5: We will work together to enhance our effectiveness in implementing the Principles.*
>
> *Principle 6: We will each report on our activities and progress towards implementing the Principles.*

Sustainable investing (49news) encompasses a menu of strategies that can be combined. Some common ones are screening (for example exclusionary or alternatively best ESG performance), sustainability themed investment (such as focus on renewable energy), impact investing (looking for companies that make a positive impact on an ESG issue). ESG integration (using ESG factors in fundamental analysis) and active ownership (engaging deeply with portfolio companies).

In 2015, to ensure sustainability, the UN set Sustainable Development Goals (45website). The **2030** Agenda for Sustainable Development provides a shared blueprint for peace and prosperity for people and the planet, now and into the future. The 17 **Sustainable Development Goals (SDGs)** are an urgent call for action by all countries - developed and developing - in a global partnership. They recognize that ending **poverty** and other deprivations must go hand-in-hand with strategies that improve **health and education, reduce inequality,** and **spur economic growth** – all while tackling **climate change** and working to **preserve our oceans and forests.**

## Sustainability initiatives

An operating period of decades is needed to calculate the Return on Investment (RoI) for major capital expenditure. Process and power plant investments are designed to operate in excess of 30 years. In addition, active measures need to be taken to continually reduce any negative impacts on the environment.

Intellectual property investment is derived from ideas and technology. Development of learning organization methods is particularly important for staff development and succession. This ensures technological advancement to stay ahead of the competition.

Naturally, sustainable natural resources are required for a sustainable business. Customers are becoming more aware of the sources of products they buy and whether these are damaging or depleting the natural environment. ***Box 1.13*** discusses the perception that bio diesel is a sustainable source of energy that will help to save the environment.

### Box 1.13: Are we really saving the environment by buying bio diesel?

*In many countries, service stations sell diesel with up to 10% produced from non-fossil materials, mostly palm oil. Malaysia and Indonesia are the biggest palm oil producers in the world and log vast tracts of rain forest to create palm oil plantations. As this is a monoculture, little else grows around or in it and **the rain forest never returns.***

*When the price of palm oil dropped a few years ago, the Indonesian government decreed that all local diesel should have a higher bio-diesel component, to increase the consumption of palm oil. And this in an oil producing country!*

*As a result of massive logging and planting of oil palms, Borneo's rainforest has declined by more than 50% since the mid-20th century (50book).*

*There is, nevertheless, an alternative. Neste, a Finnish Refining Company, has produced **renewable** diesel that is a full substitute to conventional fossil diesel **without using palm oil**. (See **Chapter 5 paragraph 'Actions being taken'**).*

Various sustainability initiatives are listed in **Chapter 5**. A number of Nordic companies are taking the lead in sustainable energy.

## Certification

Certification is a means to ensure certain standards are maintained and improved upon.

International Standards Organization (ISO)

Various standards published by the International Standards Organization (ISO) help companies with continuous improvement. Examples are ISO 9001 Quality Management, ISO 14001 Environmental Management and ISO 45000 Occupational Health and Safety Management.

ISO 9000 is a set of international standards applicable to quality management and quality assurance that have been developed to help companies effectively document the quality system elements that need to be implemented to maintain an efficient quality system. They are not specific to any one industry and can be applied to organizations of any size. ISO 9000 can help a company satisfy its customers, meet regulatory requirements, and achieve continual improvement. However, it should be considered a first step, or the base level of a quality system, and not a complete guarantee of quality.

ISO 14000 is a series of environmental management standards that provide a guideline or framework for organizations that need to systematize and improve their environmental management efforts.

ISO 45001 is an international standard for management systems of occupational safety and health. The goal of ISO 45001 is the reduction of occupational injuries and diseases.

Supply chain: International Sustainability and Carbon Certification (ISCC) (51website)

Optimizing the supply chain management for sustainability is critical. Traceability throughout the supply chain enables each player to source sustainable products from any certificate holder.

Key components of a supply chain are:

Point of origin-> Collecting point-> Trader/ storage-> Processing unit-> Final product.

International Sustainability and Carbon Certification (ISCC) is an excellent mechanism to achieve sustainability. ISCC's objective is to contribute to the implementation of environmentally, socially and economically sustainable production and use of various sources of biomass in global supply chains. ISCC's primary activity is to establish traceability in global supply chains. Other activities include implementing social and ecological sustainability criteria, monitoring deforestation-free supply chains, avoiding conversion of biodiverse grassland and calculating and reducing GHG emissions.

The American National Standards Institute (ANSI) is the appointed third party to accredit certification bodies that conduct ISCC certification.

Austrian Chemical Company, Borealis (52news), has started to produce polypropylene (PP) based on Neste produced renewable feedstock in its production facilities in Belgium. The Belgian plants were recently awarded with ISCC Plus certification by the International Sustainability and Carbon Certification (ISCC) organization for its renewable PP.

Forestry certification

Forestry certification is becoming the norm where responsible governments and purchasers require evidence of the source of the wood products.

Sustainable forest management is the management of forests according to the principles of sustainable development which aims to keep the balance between three main pillars: ecological, economic and socio-cultural.

Forest lands are classified as either commercial or noncommercial. Commercial forest lands are "capable" and "available" for growing trees for harvest. In addition to providing fiber for wood and paper, they provide recreation, wildlife habitat, and watershed protection.

Certification alternatives are outlined in **Box 1.14**.

> Box 1.14: What is the difference between FSC® and PEFC? (55wiki)
>
> *The Forestry Stewardship Certificate (FSC) (53website) is awarded by the Forest Stewardship Council® (FSC) . The mark means that the wood or paper you are buying comes from responsibly managed forests. To become FSC certified, forest operations must meet demanding environmental, social and economic requirements, as confirmed by an independent third party such as SCS Global Services.*
>
> *The Programme for the Endorsement of Forest Certification (PEFC) (54website) is an international, non-profit, non-governmental organization which promotes sustainable forest management through independent third party certification. It is considered the certification system of choice for small forest owners.*
>
> *FSC® sets specific standards that have to be met by suppliers in the timber trade. PEFC is an umbrella brand incorporating different national certification schemes. The end goals of both bodies are the same.*

Carbon neutral certification (56website)

Carbon neutral certification demonstrates an organization's commitment to sustainability, reduction of carbon emissions and the support of environmental projects. Certification is based on PAS 2060, the internationally recognized specification for carbon neutrality. It sets out requirements for quantification, reduction and offsetting of greenhouse gas (GHG) emissions.

BCorps certification (78website)

Certified B Corporations are businesses that meet the highest standards of verified social and environmental performance, public transparency, and legal accountability in a bid to balance the goals of profit and purpose. The combination of third-party validation, public transparency, and legal accountability help Certified B Corporations build trust and value. B Corp Certification is administered by the non-profit B Lab.

## Government guidance on sustainability

Ideally, governments should establish sustainable development policies to attract investments. Some governments offer subsidies for limited periods to introduce sustainable alternatives to current polluting industries. **Box 6.7** gives an example of a state subsidy to promote wind energy. However, there are cases where governments promote extreme polluting practices (57movie).

> *"US gas 'fracking' does not require either EPA approval or compliance with their clean water regulations, as the chemicals used are proprietary. This despite extremely serious environmental damage being wreaked in large parts of the US."*

For exploitation of natural resources, governments issue licenses to companies. Oil and Gas licenses entail Production Sharing Agreements between the government and oil companies (see **Box 1.15**). Surface mining licenses or concessions often require that the surface be reinstated. Logging concessions (58website) are awarded for exploitation of natural forests and often entail a FLEGT (Forest, Law, Enforcement, Governance and Trade) license and Voluntary Partnership Agreement (see **Box 1.16**).

### Box 1.15: Production Sharing Agreements (PSAs)

*PSAs are contracts signed between a government (usually through the State Petroleum Company) and an experienced upstream technology company (the contractor). PSAs are one of the most popular of the various petroleum fiscal arrangements and can be beneficial to governments that lack the expertise and/or capital to develop their resources and wish to attract foreign* companies to do so. They can be very profitable agreements for the oil companies involved, but often involve considerable risk.

## Box 1.16: Voluntary Partnership Agreements (VPAs) and FLEGT (Forest, Law, Enforcement, Governance and Trade) licenses
(59website)

*Indonesia, one of the world's largest timber exporters, has become the first country to issue FLEGT (Forest, Law, Enforcement, Governance and Trade) licences. This marks a significant step in the fight towards preventing trade in illegally harvested timber and timber products. FLEGT licenses are a key component of Indonesia's Voluntary Partnership Agreement (VPA) with the EU, which encourages governance and legal reforms so that the legality of timber can be guaranteed.*

*Indonesian timber products covered by the FLEGT licensing scheme must now have a FLEGT licence to access the EU market, and EU companies buying timber with a valid FLEGT licence can consider it legal for the sake of compliance with the EU Timber Regulation.*

*Beyond Indonesia, five other countries are implementing VPAs (including Ghana, Liberia and the Republic of Congo) and nine others are in negotiations with the EU (including Gabon and Ivory Coast).*

Emily Unwin, Head of ClientEarth's Climate and Forests Programme (60website), said:

*"This is a major step forward. By agreeing to issue FLEGT licences to all timber product exports, not just those going to the EU, Indonesia has committed to improving forest governance in a significant way. The scheme should help preserve forests, protect community rights and ensure illegal timber harvesting is prevented."*

Research is necessary to identify effective policies for forest and landscape management. **Box 1.17** outlines the work of a key researcher.

## Box 1.17: Center for International Forestry Research (CIFOR)
(61website)

*CIFOR is a non-profit, scientific institution that conducts research on the most pressing challenges of forest and landscape management around the world. Using a global, multidisciplinary approach, they aim to improve human well-*

*being, protect the environment, and increase equity. To do so, they conduct innovative research, develop partners' capacity, and actively engage in dialogue with all stakeholders to inform policies and practices that affect forests and people.*

## 3 Ethical leadership and eradication of corruption

Lack of ethics, bribery and corruption go hand in hand.

Ethics is concerned with those values and morals that an individual or a society finds desirable and/or appropriate. A leader's choices will be influenced by his/her moral development.

Ethical leadership is directed by respect for ethical beliefs and values, and for the dignity and rights of others. It relates to concepts such as trust, honesty, consideration and fairness. It therefore needs to be cultivated and upheld. Incumbents often tend to become corrupt over time, so the period that leaders serve should, ideally, be limited. Succession planning is essential for a seamless transfer of power. Roles of leaders need to be divided to ensure checks and balances on power. For example, the roles of chairman and CEO should be separate.

Ethical leadership is required for the long term survival of a company. Unscrupulous public companies might offer good profits in the short term, but are likely to be exposed over time. Having said this, private companies could operate for long periods in spite of poor ethics.

Corruption is defined as *'the abuse of entrusted power for private gain'*. It is a disease and is detrimental to the existence of a company over the long term. Weak governments and predatory international companies and individuals cause environmental pollution and grief to the communities in which they operate. Where possible, corrupt practices need to be 'nipped in the bud'.

South Korea addressed a situation of corruption quickly and efficiently by removing their president, her adviser and a number of companies which were involved in a corruption scandal (62news).

Bribery is the act of giving or receiving something of value in exchange for some kind of influence or action in return, that the recipient would otherwise not offer.

Many types of payments or favors can constitute bribes: tips, gifts, favours, discounts, free tickets, free food, free adverts, free trips, kickbacks, inflated sale of objects or property, lucrative contracts, donations, campaign contributions, sponsorships etc. Companies that offer bribes get an unfair advantage over other companies. These companies call it a 'competitive edge'. As a result, honest companies decline to quote, prices go up, projects may be delayed and the work may not be done satisfactorily, if at all.

> *South African State Owned Enterprise (SOE), Eskom, is building two mega power stations (9,600MW of power), Kusile and Medupe. Both projects have been plagued by corruption. Eskom has overspent R52.2bn on the two projects. At the time of writing, Kusile is 5 years behind schedule while Medupe is 6 years behind. (63news)*

## Fundamentals (41book)

The fundamentals of good governance include fairness, transparency and independence.

Ethics can be summarized as:

1. legality of an action
2. fairness of an action
3. personal feeling of well-being with respect to an action.

Business ethics are an integral part of both governance and performance, and can be managed on two levels:

- At the macro level, there are issues about the role of business in society. The broad ethical stance of an organization is managed within a Corporate Social Responsibility (CSR) policy.
- At the individual level, business ethics is about the behavior and actions of the people within organizations and is managed at staff

level.

Zero conflict of interest and a high standard of ethics are essential for good governance.

Similarly, fair tendering processes are crucial to retain contractor interest and permit them a fair profit so as to stay in business.

Royal Dutch Shell has had its ups and downs: the Wiwa case in Nigeria and the false reporting of oil reserves (13news)(14news) are a few examples. Texaco's unethical activities in Ecuador (64movie), where the environment was destroyed and the community's health suffered, was never resolved.

Refusing to be involved with bribery and corruption to obtain contracts or for maintaining a business presence in a country demonstrates high ethical standards. After publishing the BP Code of Ethics, BP refused to pay the Indonesian army for protection of its LNG operations. In contrast, the American PT Freeport paid millions to the Indonesian army to protect its Grasberg gold and copper mine (65news).

## Statement of ethics

A clear 'ethics statement' is now mandatory for any progressive public company. An outline for development is shown in **Box 1.18**.

## Box 1.18: Key steps for development and implementation of effective **codes of ethical practice**

> 1. They must be directly **set in the context of the Mission and Values of the organization**. This is critical for presenting the code as a coherent, strategic document and serves to trigger mutual endorsement. This can be neglected when legal concerns take priority in the development of codes.
> 2. The code should **emanate from the board**, be embraced willingly and practically by all of the senior leadership team and be introduced/endorsed by the Chairman and Chief Executive Officer.
> 3. In both development and implementation it is vital that the code is

***widely circulated*** within the organization, and externally, and that it is presented in an appropriate form, simplified or translated as necessary. Recipients need to commit to the code in theory and in practice; they can only do so if they are able to engage with it.

4. The code must **contain sanctions for non-compliance and must be seen to be enforced**. Breaches must be reported to the Board and to the internal and external public as appropriate.

5. Implementation of the code must be **supported by interactive training** to ensure that the code is understood and to measure effectively its impact on recipient values, attitudes and behaviors.

6. To support the effective operation of the code, **senior management must put in place practical, fair, impartial and, where necessary, confidential procedures for providing advice** to those with the task of making decisions in relation to the code.

7. Similar **supportive** provisions are required, in practice, with regard to the **process for raising an issue for investigation linked to the provisions of the code**.

## Board of Directors: The 'fulcrum' of business performance

(48book)

> *"the board should be able to exercise objective judgment on corporate affairs independent, in particular, from management."* OECD statement

The Board is responsible for making appropriate decisions to take the company forward by using information drawn from both the external environment and business operations.

The Board must lead in the formulation of policy and approve the Company Policy Statement (Mission, Vision, Values etc.). It must also undertake strategic thinking to determine the focus within the company's restrained resources (financial and human capital) and thus approve strategies for achieving the Company Policy, all within the **company's risk appetite** which has also been determined by the Board. This is recorded in the Corporate Risk Statement.

Basic requirements for Boards include:

1. Boards must be seen to operate 'independently' of the management of the company. So the role of non-executive directors is increased.
2. Board members should be suitably competent to scrutinize the activities of managers. The collective experience of the board, its training and the information at its disposal, are crucially important.
3. Directors should ensure that they have sufficient time to be effective. Limitations on the number of directorships that an individual can hold are therefore an important consideration.

*"..respect, trust, 'constructive friction' between board members, fluidity of roles, individual as well as collective responsibility, and the evaluation of individual director and collective board performance.." (83book).*

Company laws generally impose fiduciary duties/ duties of loyalty and good faith, and other duties on Board Members including:

1. Act in good faith in the best interests of the business
2. Comply with confidentiality requirements
3. Ensure they have adequate skills
4. Avoid conflicts of interest
5. Not exploit their position for their own gain and
6. Make full and fair disclosure of all material matters to the board and/or the shareholders.

The Board must also approve conflict of interest and ethics corporate statements.

Guidelines for effective corporate governance related to board appointments are listed in **Box 1.19**.

## Box 1.19: Guidelines for effective corporate governance

*1. The Chairman should be independent and not an executive of the Company.*
*2. The Chief Executive should not be Chairman, and the roles of each should be separate.*
3. At least half of the Board should be non-executives, excluding the Chairman.
4. Full-time executives should limit their work with other companies.
5. A senior independent director should be appointed the champion of shareholders' concerns.
6. Non-executives should serve no more than two 3-year consecutive terms.
7. Non-executives should receive no further remuneration, pensions, or share options other than their standard fees.
8. A nominations committee should monitor time input of non-executives.
9. Non-executives should not simultaneously sit on the audit, nomination, and remuneration committees.
10. Non-executives and boards should be suitably trained and regularly appraised.
11. Non-executives should be drawn more from non-board members of other businesses and the public sector.
12. There should be total transparency on appointments, training, and attendance of non-executives.

## Corruption legislation and enforcement

How do we eliminate corrupt practices? A tightening up of control in a few areas is discussed. The fight against corruption requires the passing of laws and their implementation, adherence to proper standards and the exposure of corrupt practices, by whistleblowers and investigative journalists, for example.

In the 70s, Lockheed Aircraft Company (67news) was found to be bribing members of the German, Dutch, Italian and Japanese Governments to obtain contracts for supply of its aircraft. (It is interesting to note that one of their aircraft, the Lockheed Hercules C130, is regarded by some as the best military transporter of the 20$^{th}$ century and has been the longest continuously produced military aircraft for over 60 years).

The U.S. introduced the Foreign Corrupt Practices Act (FCPA) in 1977 to address bribery of foreign officials. This legislation dominated international anti-corruption enforcement until around 2010 when other countries began introducing broader and more robust legislation, notably the United Kingdom Bribery Act 2010. In recent years, cooperation in enforcement action between countries has increased, for example, the Swiss and Singapore authorities have taken action against company officials linked to the Malaysian 1MDB scandal (see **Chapter 3**).

The control of stock exchanges around the world has been progressively improved. The US Stock Exchange Commission (SEC) has fined a number of companies for corrupt practices (see **Chapter 3**). Recently the Nigerian SEC (80) has ordered the dismissal of an entire Board of directors.

> *"The **SEC** promotes full public disclosure, protects investors against fraudulent and manipulative practices in the market, and monitors corporate takeover actions in the United States."* (82website)

Legislation is the easy part, but enforcement is critical. Enforcement requires strong independent State Watch Dogs.

Singapore, under the leadership of Prime Minister Lee, advocated a total intolerance to corruption enforced by a Corrupt Practices Investigation Bureau (CPIB), which has an excellent track record.

In contrast, Malaysia's Anti Corruption Commission (MACC) (83book) failed to address the Prime Minister Razak Najib's corrupt activities when he shuffled the members of the MACC.

In South Africa, success in the fight against corruption in government depends on the Office of the Public Protector (68wiki) (69website), which is one of six independent state institutions set up by the country's Constitution to support and defend democracy. The Public Protector is appointed by the President for a non-renewable period of 7 years. The previous Public Protector, Thuli Madonsela, uncovered 'State Capture' by 3 Indian brothers, the Guptas, who received favorable contracts from State Operated Enterprises and even influenced the appointment of Cabinet Ministers, due their influence over the President, Jacob Zuma (see **Chapter 3**). The current Public Protector, Busisiwe Mkhwebane, appointed by Zuma, has been implicated with connections to the Guptas by HSBC (see **Chapter 3 paragraph 'State Capture'**).

**As can be seen in the examples of South Africa and Malaysia, the system fails when corruption has spread to the very top and so it is very difficult to regain a trustworthy reputation.** Checks and balances are required between the executive/administration, legislature and judiciary arms of government. The Brexit fiasco in the UK demonstrated how the executive could be overruled by the judiciary when Prime Minister Boris Johnson unsuccessfully attempted to suspend parliament.

## Recommendations to combat corruption (70website)

Transparency International recommend the following in the attempt to stop corruption.

1. Strengthen institutions and preserve checks and balances
2. Close the implementation gap between anti-corruption legislation, practice and enforcement
3. Empower citizens to speak out and hold governments accountable
4. Protect press freedoms so no journalist has to fear for their lives when reporting on corruption.
5. Encourage constructive whistleblowing.

## International best practices

International best practices are used in external verification processes to measure and ensure that a program of bribery prevention works and is consistent with international standards.

Organization for Economic Co-operation and Development (OECD) Anti-Bribery Convention. (71wiki)

The OECD Anti-Bribery Convention (officially Convention on Combating Bribery of Foreign Public Officials in International Business Transactions) is an anti-corruption convention of OECD aimed at reducing political corruption and corporate crime in developing countries, by encouraging sanctions against bribery in international business transactions carried out by companies based in the Convention member countries. Its goal is to create a truly level playing field in today's international business environment. The Convention requires adherents to criminalize acts of offering or giving bribes, but not of soliciting or receiving bribes.

G20 Anti Corruption Working Group (ACWG) (72website)

Established in 2010 and an active partner of the OECD, the ACWG facilitates cooperation in raising the standards of transparency and accountability as well as contributing to the global fight against corruption.

Transparency International (TI) Business Principles for Countering Bribery (73website)

The Business Principles for Countering Bribery provide a framework for companies to develop comprehensive anti-bribery programs. Companies are encouraged to consider using the business principles as a starting point for developing their own anti-bribery programs or to benchmark existing ones.

## International Standards Organization (ISO) Standards

ISO 20000:2010 Guidance on social responsibility (74website)

ISO 26000:2010 is intended to assist organizations in contributing to sustainable development. It is intended to encourage them to go beyond legal compliance, recognizing that compliance with law is a fundamental duty of any organization and an essential part of their social responsibility.

ISO 37001:2016 Anti-bribery management systems (75website)

The ISO introduced an international anti-bribery management system standard in 2016. It is designed to help an organization establish, implement, maintain, and improve an anti-bribery compliance program. It includes a series of measures and controls that represent global anti-bribery good practice.

The organization adopting this standard must implement a series of measures and controls in a reasonable and proportionate manner to help prevent, detect, and deal with bribery, including:

1. Anti-bribery policy
2. Management leadership, commitment and responsibility
3. Personnel controls and training
4. Risk assessments
5. Due diligence on projects and business associates
6. Financial, commercial and contractual controls
7. Reporting, monitoring, investigation and review
8. Corrective action and continual improvement.

Benefits include:

• Minimum requirements and supporting guidance for implementing or benchmarking an anti-bribery management system
• Assurance to management, investors, employees, customers, and other stakeholders that an organization is taking reasonable steps to prevent bribery
• Evidence in the event of an investigation that an organization has taken reasonable steps to prevent bribery.

## 4 Expecting the unexpected

Expecting the unexpected is an art as well as an ability to undertake analysis. It is an integral part of the annual corporate strategic planning process where future plans are linked to anticipated budgets.

## Scenario planning

Scenario Planning is used in conjunction with other models to develop alternative views of the future. Scenarios are detailed and plausible views of how the business environment of an organization might develop in the future, based on key drivers for change (about which there are different levels of uncertainty).

Scenarios have been used for years as a tool to help organizations anticipate and adapt to unpredictable and uncontrollable contexts. They are a set of carefully constructed stories about how the world around us might unfold. They are internally consistent hypotheses about the future that are simultaneously relevant, challenging, plausible and clear.

Scenario work can be a powerful means for a diverse group of stakeholder leaders to collectively consider their possible futures and consider ways in which they can contribute to creating the best one possible. Scenario processes are effective because they have an unusual combination of characteristics: They are informal and noncommittal, logical and challenging, inclusive and holistic, collective and constructive, and choice-eliciting and generative.

Scenarios exhibit four key characteristics (76website):

1. Scenario building is a tool to facilitate strategic decision-making by top management
2. Scenario thinking is used as an advocacy tool, whether for an individual organization or for a wider system
3. Scenario based planning is a tool for generating awareness with a view to building understanding and perhaps eventually consensus
4. Scenarios are exercises in intellectual enquiry.

Scenario processes can produce four types of results:

1. Systemic insights and understandings about what is happening and what might happen, and what these factors imply

2. Stronger relationships and alliances among leaders from across organizations and sectors

3. Clearer intentions and commitments as to what these leaders need to do

4. As a result of these insights, relationships, and intentions, innovative initiatives and actions to co-create a better future.

## Black Swans

In risk terminology, a Black Swan is an "unknown unknown," which is an unforeseen event. It goes without saying that there are events that cannot be predicted, although scenario planning may include events such as tsunamis, but might not cater for its magnitude or timing.

Can one mitigate the impact of a Black Swan? When the 2011 Japanese Tsunami occurred, Tokyo Electricity Power Company (TEPCO) appeared to have taken insufficient mitigation steps (see **Box 3.2**).

Some of the events in **Box 4.10** could be regarded as Black Swans.

## Preparedness

We don't know what is on the horizon. We can learn from the oil and nuclear industries which have developed Incident Preparedness and Operational Continuity Management (IPOCM) protocols, to the point of being formed into two ISO standards (81website). Both governments and companies need to recognize disasters quickly, and respond immediately to prevent escalation. In addition, a well considered and structured approach must be established to deal with unforeseen incidences.

When the 2019 Covid pandemic occurred, the USA was unprepared, while some countries had structured strategies that were immediately put into action to minimize damage to their economies.

## Environmental tipping point

A number of factors occurring simultaneously could cause an environmental tipping point to be reached. These include melting of the permafrost, loss of snow reflection, peat bogs drying out, mangroves being filled in, coral bleaching, glaciers carving (see **Chapter 5**). **Already the bleaching of 29% of coral on the Great Barrier Reef has occurred over 2 years.**

## Asteroid collision with Earth (77news)

Possible collisions could take place in 2022 and 2027. However, the probability is small.

## Viruses

After the SARS virus epidemic in 2003, some countries set up processes in anticipation of another virus scare where others did little or nothing.

Ebola sprouted up in the east of the Democratic Republic of the Congo (DRC) and hopefully has been suppressed. This is the DRC's 11th outbreak of Ebola since the virus was first discovered in the country in 1976.

The 2019-20 outbreak of the Covid-19 virus has had a major effect on businesses and the world economy.

## Conclusion

Businesses need to move away from short-term profit motives and replace those values with more sustainable long-term business models. They must consider their staff, the communities in which they operate and the surrounding natural environment, and they need direction from shareholders in this quest.

# CHAPTER 2: INFLUENCERS

## Introduction

Businesses can be influenced by many different parties, with stakeholders (*"any group without whose support the organization would cease to exist"*) having the largest effect. Influencers such as shareholders, customers, suppliers, communities, governments, lobbyists, protectors and disruptors can have a major impact on a company's direction and profitability.

In this chapter I take a deeper look at each type of influencer and their effects, whether positive or negative, on business and the economy.

## Shareholders

Shareholders are the watch dogs of companies. Besides making profits for their shareholders, companies must focus on other 'shareholder wishes' such as reduction of environmental emissions and optimization of energy use.

### Institutional investors (1news)(2news)

> "Using a broad measure, there was global **sustainable investment** of $30.1 trillion across the world's five major markets at the end of 2018, more than a quarter of all assets under management globally. That compares with $22.8 trillion in 2016." (131news)

Institutional investors, such as pension funds and insurance companies, can have a considerable influence over the allocation of their investments. This influence will hopefully play a greater role in the future by ensuring investments in more ethical companies. More and more institutional investors are insisting that Environment, Social and Governance (ESG) issues be addressed, with specific emphasis on climate change. Private companies, however, are not obliged to disclose their activities to anyone other than the tax man. Some examples of Institutional Investor groups include:

*Climate Action 100+*

*Climate Action 100+ is a five-year initiative led by investors to systemically engage the world's largest corporate greenhouse gas emitters to drive the transition to clean energy, and help achieve the goals of the Paris Agreement. To date, 310 investors with more than USD $32 trillion in assets under management have signed on to the initiative.*

*Institutional Investors Group on Climate Change (IIGCC)*

*The IIGCC is a network of nearly 150 members, including nine of the 10 largest pension funds and asset managers in Europe, who represent over €21 Trillion in assets and take a pro-active approach to managing risks and opportunities related to climate change. IIGCC offers opportunities to deepen investor understanding of climate risks and opportunities to ensure that these are reflected in investment practices which will preserve and enhance long-term investment value.*

Shell shareholders recently agreed with Shell directors to focus on the requirements of the Paris Accord. I give an outline of this historical agreement in **Box 2.1.**

### Box 2.1: Royal Dutch Shell Shareholder Agreement (edited)

[3website]

### Joint statement between institutional investors on behalf of Climate Action 100+ and Royal Dutch Shell plc. (Shell)

*"... we, the Institutional Investors and Shell, are pleased to jointly announce the steps below that Shell has decided to take in order to demonstrate further industry leadership and alignment with the goals of the Paris Agreement on climate change. As Institutional Investors, we are strongly supportive of the company in taking these important steps."*

*Summary headings*

1. Public short-term Net Carbon Footprint targets
2. Targets linked to remuneration
3. Review of progress
4. Alignment with the Taskforce of Climate related Financial Disclosure (TCFD) recommendations
5. Corporate climate lobbying.

## Asset managers

Asset managers manage the investments of institutional investors.

Blackrock (4website), one of the top three investment management companies in the world, has announced a focus on sustainability for future institutional investments. It has also joined the Climate Action 100+ pressure group. The other two, Vanguard (5) and State Street (6), have come under fire for climate-damaging investments, particularly in fossil fuels. Vanguard has an especially poor record for supporting climate-related shareholder resolutions.

BNP Paribas (108news) Asset Management committed to align all of its portfolios with the "well below 2°C" Paris Agreement a year after it was signed in 2015. It set itself an overall target of achieving these goals by 2025, but started by excluding coal from its portfolios from January 2020.

Morningstar (109news), a financial services firm, describes the trend towards ESG funds as "record-shattering" with $120bn invested during 2019, with the trend accelerating towards the end of the year.

Europe's largest asset manager Amundi (109news), with Assets under Management (AuM) of €1.65tn ($1.78tn), announced in mid-February 2020 that it was backing a shareholder motion to stop UK bank Barclays from offering loans to all fossil fuel companies.

## Primary stakeholders other than shareholders

Besides shareholders, primary stakeholders include staff, suppliers, customers and the communities in which they operate.

### Staff

Staff interface with suppliers and customers and ensure the required quality and quantity of product. They are key to the success of the company.

### Suppliers

Suppliers determine the quality of the raw materials and in many cases need to be nurtured over long periods to ensure that the desired quality is maintained. As large businesses have a big influence over their suppliers, ethical business practices to ensure congenial long term relationships are required throughout the supply chain. (See **Box 4.2 Fairtrade** and **Box 4.3 Woolworths**).

### Customers

Customers decide on what and where to buy and, ideally, as customers, we should choose suppliers based on ethical standards. Our social conscience should dictate that we refuse to buy products from corrupt and polluting companies, but this is easier said than done of course. Few of us have the wherewithal to do the research required, although there are some retailers who are doing some of the work for us. IKEA, as a customer, has encouraged responsible coffee growing, promoted ethical clothes manufacture (not using child labour) and addressed the plastic waste issue. There are other customer initiatives to reduce carbon emissions and damage to the environment such as:

1. Palm oil - avoiding bio-diesel, see **Box 1.13**

2. Tropical hard wood - choosing wood from approved sustainable forests, see **Box 1.16**

3.Transport - using electric cars instead of petrol engined cars.

As consumers, we should focus on reducing our carbon footprint and, since energy consumption for day-to-day living is such a significant contributor, it is well within our means to do so.

Customers can also rate companies using social media. This is now a significant influence on the success of many companies.

## Communities

Communities that are directly affected by a company's operations, include company staff as well as various service industries.

# Governments

A business can only survive with the goodwill of the government and people of the country in which it operates. It is often difficult for businesses to survive during political instability unless there are corrupt practices in play. Even though the people tend to rise up when these practices become unbearable, it is often the case that suppression of the masses continues.

**It stands to reason that some degree of political stability is desirable to encourage companies to invest.**

Commercial laws need to be fair to encourage the formation of businesses and for the equitable purchase and supply of goods and services. Infrastructure such as electricity, water, telecommunications and transport need to be developed to support companies as well as attract investment.

## Democracy and investment

A stable democratic system is regarded as a positive indicator for long term investment. According to Freedom House (8website), 113 countries have seen a decline in their democracy scores since 2006.

Countries are rated from zero for 'not free' to 100 for 'totally free', based on:

1. Free and fair elections.
2. Strong and independent institutions.
3. Political rights, for example, the right to protest.
4. Civil rights, such as the access to a fair trial.

The Corruption Perception Index probably gives a fairer indication how well democracy is working (9website).

> "Corruption is much more likely to flourish where democratic foundations are weak and, as we have seen in many countries, where undemocratic and populist politicians can use it to their advantage." (9website)

Progress towards democracy with minimal corruption has been slow in Africa. The Mo Ibrahim Prize has given an incentive for heads of state to democratically transfer power (see **Box 2.2**).

### Box 2.2: Mo Ibrahim Prize (10wiki)

*The Mo Ibrahim Prize was established by the Mo Ibrahim Foundation in 2007. This award celebrates excellence in African leadership and is awarded to a former Executive Head of State or Government by an independent Prize Committee composed of eminent figures, including two Nobel Laureates.*

*With a US$5 million initial payment, plus $200,000 a year for life, the prize is believed to be the world's largest, exceeding the $1.3m Nobel Peace Prize.*

*The last person to receive this prize was Ellen Johnson Sirleaf, from Liberia, in 2017. Nelson Mandela of South Africa was an honorary recipient in 2007.*

## Governance

> '*Governance* encompasses the system by which an organisation is controlled and operates, and the mechanisms by which it, and its people, are held to account.' (136website)

Governance is a set of guidelines which proposes how to drive an organization forward whilst keeping it under prudent control. Countries need stable governance (12website) that supports broad-based growth throughout society. Key elements are:

1. An independent and credible central bank which acts to keep inflation in check and mitigates currency crises,

2. A fair and efficient tax regime which redistributes income fairly among all members of society, supports social development (including education and health care) and sets out a clear and stable taxation policy applicable to all businesses, ▪

3. An appropriate degree of intellectual property rights protection to ensure that individuals and businesses are rewarded for their innovations and entrepreneurship, ▪

4. Effective environmental regulations which promote sustainable economic growth and

5. The cultivation of high levels of trust between government, business and the general population.

Corporate governance codes have been established in many countries, some of which I've described in **Chapter 1**. The Ibrahim Foundation has established a Governance Index (Ibrahim Index of African Governance - IIAG) for Africa (11news) whilst in Asia, ASEAN has initiated the Corporate Governance Scorecard (see **Box 1.7**).

## Sustainability

Economic wellbeing is a prerequisite for any country's long term economic sustainability. (Wellbeing is defined as the state of being comfortable, healthy, or happy.) Sadly, the current trend is to measure economic growth based solely on GDP and this practice is proving very destructive (7audio). Social progress should also be a dominant indicator (see **Chapter 4 paragraph 'Existing institutional performance indicators - Social Progress'**). For instance, the Happy Planet Index (HPI) (125website) ranks countries based on their environmental impact as well as the health and happiness of their citizens. This index should be promoted and incorporated into business models and policies. Countries that maximize investment in health and education, and minimize investment in the military, have a higher HPI resulting in a far more sustainable economy. The Universal Human Rights Index (UHRI) is a good indicator to show progress related to the 17 UN Sustainable Development Goals (137website).

Costa Rica has been ranked no 1 according to the Happy Planet Index (HPI) (13news). Costa Rica does not have a standing army and so spends tax money on more important things such as health and education. A Costa Rican has an ecological footprint of one-fourth of the average American (US).

Wise 'value addition' (conversion from raw materials to usable products) of a country's natural resources with the minimal environmental impact is necessary, as well as the use of the sun and wind for power generation.

Qatar, one of the biggest Liquified Natural Gas (LNG) exporters in the world, uses the revenue from oil and gas for the benefit of the Qatari people and to invest in the Qatar Sovereign Wealth Fund. In the past 15 years, their education system has had a major overhaul and a number of prestigious American universities have established satellite campuses in the country. Government health centres and hospitals have been established, while infrastructure such as potable water, effluent and roads have been upgraded and expanded, and a new metro rail system established. Qatar spends a minimal amount on defense due to an agreement with the US which sanctions a US military base in the country. On the downside, Qatar is ranked 3rd worst in HPI for its eco-footprint (138news).

On the other hand, Nigeria has high expenditure on military while very little of the country's oil revenue is used for the benefit of the people i.e. there is little 'value addition' of the oil and gas that the country produces. They have even had to import of petrol at times, in spite of being Africa's largest producer of oil and the sixth largest in the world. Only 40% of Nigeria's population is connected to the energy grid while power supply difficulties are experienced around 60% of the time. Nigeria ranks 95th on the HPI, although conflict has intensified in recent times and this ranking may no longer be reliable.

## Electrical energy production

Electrical energy production is a major factor in driving the economy of any country and in many cases is both owned and managed by a State Owned Enterprise (SOE). The cost of electricity production is a critical factor in the growth of a state's economy. Regrettably, the prime cause of global warming is from this production of energy. I discuss the main offenders of $CO_2$ emissions in **Chapter 5**.

On a positive note, renewable energy sources are growing exponentially with cheaper solar and wind generation being available. Germany is the leader of solar with China ahead in wind (14book). Germany generates 42% of its power from solar and wind (132website). France has now committed to be carbon neutral by the year 2050 with 70% of its power presently generated from nuclear energy. The tiny Himalayan state of Bhutan is already carbon negative, mainly with the use of hydropower.

## Investment grade

The investment grade of a country, determined by rating agencies such as Moody, is critical for obtaining loans from international lenders such as the International Finance Corporation or regional development banks, for infrastructure projects. Investment ratings for National Bonds give another indication of the stability of a country for purposes of investment. Investment grades and ratings are discussed in **Chapter 4 paragraph Existing institutional performance measurements.**

## The resource curse

> "The resource curse, also known as the paradox of plenty, refers to the paradox that countries with an abundance of natural resources,

tend to have less economic growth, less democracy, and worse development outcomes than countries with fewer natural resources." *(15wiki)*

Oil and gas have been found in a lot of developing countries, many with weak governments. The tendency is for powerful international oil companies to exploit these governments and the people of these countries and give backhanders to those who benefit from the deals. The same is true for high value minerals. Most affected countries are in Africa (Ghana, Democratic Republic of Congo, Uganda, Republic of Congo, Angola, Mozambique and Nigeria), and in Central Asia (Kazakhstan, Uzbekistan and Azerbaijan) Angola was rumored to have lost at least a third of its GDP through misappropriation of taxes, although their new government, elected in 2017, is making changes for the better. I give an outline in **Box 2.3**.

## Box 2.3: The Angola turnaround

*2004 (19news)*

*From 1997 to 2002, unaccounted for funds amounted to some U.S.$4.22 billion. In this time, total social spending in the country-including Angolan government spending as well as public and private initiatives funded through the United Nations' Consolidated Inter-Agency Appeal-came to $4.27 billion. In effect, the Angolan government has not accounted for an amount roughly equal to the total amount spent on the humanitarian, social, health, and education needs of a population in severe distress.*

*The "Oil Diagnostic" monitoring system was set up by joint agreement of the IMF and the Angolan government starting in 2000. The Oil Diagnostic showed that billions of dollars from Sonangol, the state-owned oil company, illegally bypassed the Angolan central bank and that the government did not have any procedures in place to reconcile hundreds of millions of dollars of discrepancies in its accounting of oil revenue.*

*2017 (20website)*

*The Government of President João Lourenço elected in August 2017 is taking vigorous action to address Angola's long-standing governance issues and high corruption perceptions by:*

- *Curbing nepotism in the public administration*
- *Moving against vested interests*
- *Investigating high-level officials for potential wrongdoings*
- *Creating a stronger anti-corruption unit*
- *Approving a new Law on Competition*
- *Approving a Private Investment Law that attracts Foreign Direct Investment and*
- *Creating a one-stop shop for investors.*

***Box 2.4*** outlines the changes in oil revenues in the oil rich Arabian Gulf States.

### Box 2.4: Arabian Gulf States oil revenues

*In the 1970's the Middle East oil producing countries chose to nationalize their oil industries.*

*Saudi Arabia retained good relations with the International Oil Companies (IOCs) and flourished by implementing Technical Services Agreements (TSAs) with the IOCs. On the other hand, Iran kicked the IOCs out without compensation and consequently had sanctions applied, leading to termination of technical support. Bahrain retained a joint agreement with Chevron until 1997, when it bought them out. Qatar moved towards Joint Ventures (JVs) and Production Sharing Agreements (PSAs) with various IOCs (see **Box 1.15**). Qatar is now one of the richest countries in the world.*

## Corruption

Corrupt countries pose a high risk for investors since corrupt dealings can involve the payment of large bribes, albeit with the potential of great returns. However, due to the current trend of whistleblowing, corrupt dealings are very likely to be exposed, with the concomitant risk of negative publicity and a loss of reputation. It is interesting to note that some companies accept the high risks of operating in countries such as the Democratic Republic of Congo and South Sudan, rated as two of the most corrupt countries with Corruption Perception Indices (CPIs) of 20/100 and 13/100 respectively. Corruption is discussed in detail in ***Chapter 3 "Learning from History"***.

Major corruption often interweaves strong international companies with weak governments and corrupt politicians. The companies offer politicians financial rewards in exchange for favors, for example in the award of contracts. These contracts often involve large arms deals, resource concessions or infrastructure projects. The general population and ethical companies are adversely affected. Here are some examples of the negative effect that corruption has had:

> According to a recent study, more than one-fifth of the population of oil-and-gas rich Russia lives in poverty, while 36 per cent are at risk of poverty (16website).

> Oil-related crimes (17news) cost Nigeria US$2.8 billion in lost revenues in 2018. Shell and ENI (18news) also appeared to be implicated in related corrupt activities in 2011 and subsequent court cases have popped up in Italy, Holland and the UK.

## Banks and money laundering

Banks tend to be the first port of call for anyone wishing to loan money to set up and operate a company. When these loans are defaulted on, the banks normally take over the company or have first claim on what is left when bankruptcy is declared.

The Chinese have carried out, and funded through their banks, a number of infrastructure projects, as part of their Belt and Road Initiative. Some countries have defaulted on repayments with unfortunate results (see **Box 6.6**).

Development Banks often get involved in risky ventures of which there have been many project failures over the years. A classic example was the promotion of fishing in Lake Turkana in Kenya with funds donated by the Norwegian Government (21news).

## World Bank (22website)

The World Bank is an international financial institution that provides loans and grants to the governments of poorer countries for the purpose of pursuing capital projects. There are two institutions: the International Bank for Reconstruction and Development, and the International Development Association.

The World Bank has set two ambitious goals to reduce extreme poverty to no more than 3% by 2030, and to promote shared prosperity and greater equity in the developing world.

A number of off grid electricity projects in Africa have been successfully funded by the World Bank (24website) (25website)(126news) and the bank's Maximizing Finance for Development (MFD) (26website) approach has had many successes, but there have also been failures such as the World Bank's water project in Tanzania (23news).

## African Development Bank (27website)

The overarching objective of the African Development Bank (AfDB) Group is to spur sustainable economic development and social progress in its Regional Member Countries (RMCs), thus contributing to poverty reduction. The Bank Group achieves this objective by:
- mobilizing and allocating resources for investment in RMCs and
- providing policy advice and technical assistance to support development efforts.

## Asian Development Bank (28website)

The bank's mission is to achieve a prosperous, inclusive, resilient, and sustainable Asia and the Pacific, while sustaining their efforts to eradicate extreme poverty.

## Money laundering

Some countries' economies are highly dependent on off-shore banking and large amounts of money are often laundered through those countries with lax banking laws. It is then invested in property in other countries, where the governments don't question the source of funds used to purchase the property. This is evident in Cyprus, UK, USA, Australia and Canada. Laundered money is also often invested in expensive works of art (41news).

The conduits for money laundering include banks and tax havens. Banks implicated in various scandals include:

## Malaysian Sovereign Fund - 1MDB (see **Chapter 3**)

- Goldman Sachs (1MDB case moved to a higher court in Singapore) (29news)
- BSI (forced to close and integrated into EFG) (30news)(31news)
- Ambank (no comments from 24% shareholder ANZ bank) (32website)
- JP Morgan (Swiss Financial Market Supervisory Authority -FINMA appointed a monitor) (33news)
- Standard Chartered (fined by Monetary Authority of Singapore) (34news)
- Coutts (fined by Swiss Financial Market Supervisory Authority - FINMA and Monetary Authority of Singapore) (35news)
- Deutsche Bank (US Department of Justice probe under way) (36news)
- DBS Bank (fined in Singapore by Monetary Authority of Singapore) (37news)
- UBS Bank (fined in Singapore by Monetary Authority of Singapore) (37news)
- Falcon (closed in Singapore by Monetary Authority of Singapore) (38news)
- Rothchild (closed in Singapore) (39news)

## Russian laundromat (see **Chapter 3**)

- Moldindconbank

- Danske Bank
- Deutsche Bank
- HSBC

## Troika laundromat (see **Chapter 3**)

- Troika Dialog in Russia
- Ukio bankas in Lithuania (now defunct)
- Raiffeisen in Austria
- Commerzbank in Germany

## Mozambique (see **Chapter 3**)

- Credit Suisse

## Drug trade and terrorist financing (110movie)

- HSBC

In 2012, HSBC paid a $1.9 billion fine to avoid prosecution in US courts for accepting at least $881 million in proceeds from the sale of illegal drugs. In addition to facilitating money laundering by drug cartels, evidence was found of HSBC moving money for Saudi banks tied to terrorist groups.

In the case of Goldman Sachs and 1MDB, the bank has reached a $3.9bn (£3bn) settlement with the Malaysian government for its role in the multi-billion dollar 1MDB corruption scheme (119news).

Offshore banking havens that have been used for laundering money include:

- Channel Islands
- Bermuda
- British Virgin Islands
- **Central London** (40video)

The UK government has only recently legislated against money laundering, by enacting the following laws:

1.The Criminal Finances Act 2017 amends the Proceeds of Crime Act 2002 to expand the provisions for confiscating funds to deal with terrorist property and proceeds of tax evasion.

2.The Sanctions and Anti-Money Laundering Act 2018. In addition to sanctions regulation, the Act further makes new provisions in relation to the detection, investigation and prevention of money laundering and terrorist financing.

Auditors sometimes help to cover up fraudulent transactions. Audit companies that have been implicated in covering up shady dealings include:

- Arthur Anderson (bankrupted)
- Ernst and Young
- KPMG
- Deloitte

## Groups offering training for employment

In an ever shifting business climate, companies change focus, down size, relocate or go bankrupt. Employees continually need to prepare themselves for change by up-skilling and re-skilling, sometimes at their own expense as many companies do not offer skills enhancement unless it is directly related to profit. In some countries though, industry and trade groups offer training, often for free. In South Africa, for instance, companies pay a training levy, which is then claimed back when staff training is undertaken.

The World Economic Forum's Closing the Skills Gap Initiative (42website), launched in 2017 with a target to re-skill or up-skill 10 million workers by 2020, announced it has already secured pledges to train more than 17 million people globally, 6.4 million of whom have already been re-skilled. The Forum also announced that the Initiative is now supporting government and private partnerships aimed at promoting future skills in four countries: Argentina, India, Oman and South Africa.

## Non-Government Organizations (NGOs)

NGOs and charities in various countries have set up training facilities in local communities. For instance, in Siem Riep Cambodia, cafe management training has been developed and supported by Cambridge University Business School (43website) and a cooking school (44website) has been established at 'The Haven' restaurant, which was developed and administered by a Swiss chef and his wife.

## National Qualification Frameworks (NQFs)

The NQF qualification is based on recognized occupational standards, confers occupational competence and requires work-based assessment and/or assessment in an environment that simulates the work place.

National Qualification Frameworks (NQFs) help employers to assess the qualifications of potential employees and employees who upgrade their skills during employment. With the changing skills requirements due to loss of employment because of automation or businesses relocating, people need to continuously up-skill or change to other forms of employment. The usefulness, relevance and value of the qualification is directly linked with the workforce and skill needs of individuals, groups of learners, employers, industry and communities. Qualifications are developed collaboratively with a wide range of stakeholders, so parties can rely on the integrity of the processes used and the information provided.

The National Qualification Frameworks in the United Kingdom (45wiki) are qualifications frameworks that define and link the levels and credit values of different qualifications.

The NZQF (46website) was one of the first qualifications frameworks in the world. It is administered by The New Zealand Qualifications Authority (NZQA) and is based on clear learning outcomes – the skills, knowledge and application demonstrated to complete a specific qualification.

The South African NQF (47website) is a single integrated system comprising three coordinated qualifications sub-frameworks:

1. General and Further Education and Training (skills programs),

2. Higher Education (school and university) and

3. Trades and Occupations (college).

The South African NQF (48wiki) is depicted in *Figure 2.1.*

Figure 2.1: South African NQF

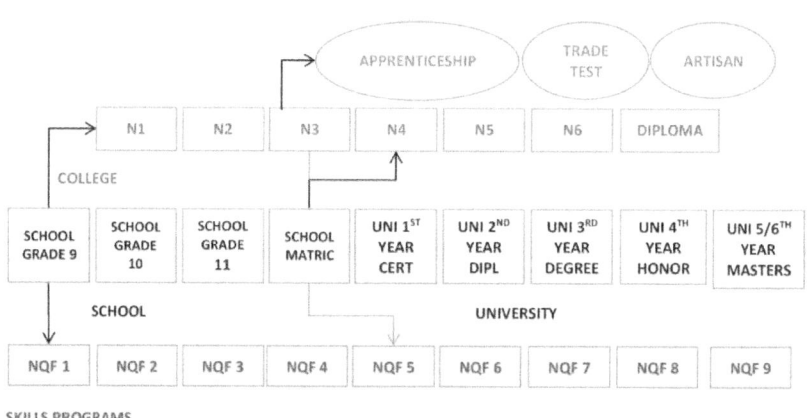

Regrettably, the USA does not have a cohesive framework for upskilling their workforce. **Box 2.5** describes the sad state of affairs in the USA with respect to worker retraining. This case emphasizes the need for continual skills upgrading and diversification to increase the potential for employment elsewhere.

## Box 2.5: Dayton Dilemma (49news)

*Dayton Ohio in the United States is where the Wright Brothers are generally credited for the world's first successful motorised flight. General Motors closed their Dayton Ohio plant when they went bankrupt in 2009. Since then a Chinese automotive glass maker has taken over the plant and offered lower salaries than GM, while resisting unionization of the plant (127movie). The company also increased automation of the glass making process resulting in a number of workers being laid off.*

## Groups that have a positive sustainable influence

These groups expose corruption and money laundering, and put pressure on countries and companies to reduce environmental degradation and abuse of people.

### Investigative reporters

Investigative journalism is a key tool for extracting factual information that is in the interests of the public. Investigative journalists often risk their lives when digging into risky and dangerous issues.

### Sarawak Report (50website)

Sarawak-born investigative journalist Clare Rewcastle-Brown founded the Sarawak Report which investigated corruption in the timber industry in Sarawak, part of the Federation of Malaysia. Her investigation was instrumental in bringing down the corrupt Prime Minister of Malaysia, Najib Razak, who is considered the biggest money launderer in history.

### Bellingcat (51website)

Elliot Higgins formed Bellingcat which is an investigative journalism website that specializes in fact-checking and open-source intelligence. Bellingcat helped to investigate the Syrian Civil War, the 2014-15 Russian military intervention in Ukraine, the downing of the Malaysian

Airlines Flight MH17 over Ukraine and the poisoning of Sergei and Yulia Skripal in the UK.

## Organized Crime and Corruption Reporting Project (OCCRP) (52website)

OCCRP provides an investigative reporting platform dedicated to improving reporting and newsroom management around the world.

## Bureau of Investigative Journalism (BIJ) (53website)

The BIJ is a nonprofit news organization based in London. It was founded in 2010 to pursue "public interest" investigations, funded through philanthropy. The BIJ identified super-bugs that were resistant to conventional antibiotics (54news).

## Centre of Investigative Journalism (55website)

The Centre for Investigative Journalism is a British independent charity providing training to journalists, researchers, producers and students in the practice and methodology of investigative journalism.

## Reporters Without Borders (56website)

Reporters Without Borders, also known under its original name Reporters Sans Frontières, is an international non-profit, non-governmental organization based in Paris that conducts political advocacy on issues relating to freedom of information and freedom of the press.

## Human rights groups

Many of these groups support the rights of indigenous people. One of the most significant abuses is seen in the logging of the tropical rain forests where indigenous tribes are very vulnerable, for example in Borneo and Amazon. These tribes have long attempted to resist companies destroying their environment. Some have succumbed to bribery in allowing loggers to log their land, and then have realized that they have lost control of the situation. Examples are described in Money Logging (133book) and the Sarawak Report (134book).

### Human Rights Watch (HRW) (57website)

HRW investigates abuse and focus on persuading governments, armed groups and businesses to change or enforce their laws, policies and practices.

### ActionAid (58website)

ActionAid is an international non-governmental organization whose primary aim is to work against poverty and injustice worldwide (see also **Chapter 1 paragraph 3 'Unethical behaviour and corrupt practices Human Rights Abuse'**).

### International Work Group for Indigenous Affairs (IWGIA) (72website)

IWGIA is a global human rights organization dedicated to promoting, protecting and defending the rights of indigenous peoples.

### Bruno Manser Fund (BMF) (71website)

The Bruno Manser Fund is committed to campaigning for the conservation of threatened tropical rainforests and the rights of the rainforest dwellers.

## Environmental groups

### Eartheasy (59website)

Eartheasy is a website that lists a number of environmental groups, each with their own agendas, and vetted for effectiveness. Eartheasy also offers information to potential donors. Recommended charities include Greenpeace Fund, Friends of the Earth, Rainforest Alliance and Ocean Conservancy.

### Masarang Foundation (60website)

Masarang's mission is "Nature conservation through collaboration with and development of the local population". Masarang seeks solutions for the most urgent global problems of our time: deforestation, biodiversity loss, climate change, poverty and underdevelopment.

### Environmental Investigation Agency (EIA) (61website)

EIA undercover investigations expose transnational wildlife crime. It works to safeguard global marine ecosystems by addressing the threats posed by plastic pollution, bycatch and commercial exploitation of whales, dolphins and porpoises. They also campaign to eliminate powerful refrigerant greenhouse gases. Their findings are used in hard-hitting reports to fight for improved governance and more effective law enforcement.

### Alliance to end Plastic Waste (62website)

*"Our Mission: Eliminate Plastic Waste In Our Environment"*

An example of an NGO's activities in Philippines is outlined in **Box 2.6**.

#### Box 2.6: Plastic Tides (63website)

*Julian Rodriguez is the founding member of Plastic Tides, an NGO focused on "combining adventure and science to fight plastic pollution".*

*"Just recently, our team at Plastic Tides Philippines teamed up with the Pasig River Rehabilitation Commission to bring in volunteers to help clean up the*

47 tributaries of the Pasig River. In just half a day, we were able to collect about 1,500 kilograms of trash! The PRRC only has about 45 people assigned to help clean up the Pasig River."

## Sea Shepherds (64video)

Sea Shepherd Conservation Society regards itself as the world's most passionate and powerful protector of ocean life, although some consider them to be eco terrorists. Your opinion might depend on whose side you're on - the ocean creatures, or the businesses that trade in them.

# Groups that have a negative influence

These groups have a seriously detrimental effect on both companies and their operating environment .

### Cyber criminals

Cyber criminals can have devastating effects on business operations. Alleged offenders include North Korean and Russian Intelligence, and China's Peoples Army.

National Oil Company Saudi Aramco suffered the worst hack in world history in 2012 (123news). In a matter of hours, 35,000 computers were partially wiped or totally destroyed and Saudi Aramco's ability to supply 10% of the world's oil was suddenly at risk.

The global Wannacry ransomware (124news) attack in May 2017 caused panic when it froze software systems and demanded ransom payments. Over 300,000 computers were infected and the UK National Health Service was overwhelmed.

Malware called 'Eternal Blue', developed by the National Security Agency (NSA), has recently been found in the City of Baltimore's computer system (65audio), where it has been used to hold the city's computer systems hostage.

## Organized crime/ mafia

Drug peddling, illegal gambling, human trafficking and prostitution are the pursuits of Organized Crime syndicates around the world. In areas where government police forces are weak, small businesses resort to paying the local mafia for protection. New York was controlled by five mafia families in the 1980s until the key players were jailed (118movie).

## Secret societies

Members of a secret societies tend to promote members of their society within the workplace.

Freemasons (69wiki) trace their origins to fraternities of stonemasons at the end of the 14th century. Ex US Presidents Lyndon Johnson and Gerald Ford were Freemasons as was Winston Churchill, the UK's wartime Prime Minister. Companies that have had Freemasons in key positions in the past include South African Mutual Insurance Company (Old Mutual), African Oxygen (Afrox) and Bahrain Petroleum Company (Bapco).

## Radicals

Radical Islamic groups are a major threat in Libya, Yemen, Iraq and Syria. Islamic State (IS) has recently lost a large part of its real estate but is still a major threat to business, especially in countries where the security forces are weak.

In Mozambique (122news), recent attacks by radicals have been made on companies working on a new LNG complex. To make things worse, the Mozambique government is employing private armies to help the army stop the rebellion. This could have long term investment implications for one of the poorest countries in the world . Many observers believe that the solution to the conflict lies in good governance, and a transparent attempt to address deep-seated economic and social grievances. This includes fair access to land, jobs, and a share of any future gas revenues. This analysis is typical of governments being blind to the needs of their people (see also **Box 2.10**).

## Intellectual property (IP) thieves

Intellectual property theft involves robbing people or companies of their ideas, inventions, and creative expressions - known as "intellectual property" - which can include everything from trade secrets and proprietary products and parts to movies, music, and software. The Russian TU-144 dubbed 'Concordski', for example, looked so like the Anglo-French Concorde that it's believed that the Soviet Union had to have engaged in industrial espionage to produce it.

A CIA agent, Rodger Ames (111wiki), provided Russia with a "huge quantity of information on United States foreign, defense and security policies" At the time of his arrest, Ames had compromised more highly-classified CIA assets than any other officer in history until Robert Hanson's arrest seven years later in 2001.

FBI agent Robert Hanson (112wiki) sold thousands of classified documents to the Russian KGB that detailed U.S. strategies in the event of nuclear war, developments in military weapons technologies, and aspects of the U.S. counterintelligence program. His espionage was described by the Department of Justice as "possibly the worst intelligence disaster in U.S. history." He was a member of Opus Dei, a secret society.

Many other spies have operated undetected. China has benefited greatly in obtaining the latest Western technology and enhancing it. What has helped is that US companies, such as Apple, have had their products manufactured in China and Taiwan. Hon Hai (Foxxcon) who make the Apple products, is now the 24th biggest company in the world (121news). Huawei (113news), a Chinese telecom company, is leading the world with 5G technology.

## Groups having both positive and negative effects

I present a sample of groups here which could affect the ongoing well being of companies either positively or negatively.

### Armies and mercenaries

With an inability to co-opt local populations in remote areas and areas of unrest, governments and companies resort to paying armies and mercenaries to protect their assets, sometimes resulting in the abuse of the local population. Communities in these areas are often vulnerable to local warlords or radical groups since their central governments don't protect them.

Academi/ Blackwater (US) (66book) is contracted to the US Government as a substitute for the US army to protect US Government assets and staff in unstable countries such as Iraq and Afghanistan. Academi has one of the biggest and most complex private military training grounds in the world.

Wagner (Russia) (67video), mainly consisting of ex Chechen war veterans, is being used by President Assad of Syria in an attempt to recapture oil installations. It also operates elsewhere, particularly in places of instability and under contract to governments or international companies.

For many years PT Freeport has been paying the Indonesian Army for protection of its Grasberg gold and copper mine (68news).

## Unions

A key element of the failure of British Motor Corporation (BMC) was the management verses union conflict. As a consequence, before establishing a plant in Sunderland, England, Nissan negotiated with the unions to put a formal grievance process in place to prevent wildcat strikes. Nissan Sunderland subsequently became the most productive car plant in Europe (73news).

In Germany, unions have representation on the Boards/ Management Committees of many companies and consequently disruption due to strikes is relatively rare.

In the USA, the longest strike in the motor vehicle industry since 1970, has just ended (74news) with a lose-lose outcome.

## Lobbyists and think tanks

Lobbyists can potentially have a positive influence on the specific interests of companies, which could be for the benefit of stakeholders as a whole or else in the interests of shareholders alone.

"K Street", in downtown Washington, DC, USA, where many lobbyists, lawyers and advocacy groups have their offices, has become a term to refer to the lobbying industry as a whole and is also widely used to symbolize the many problems that people see with the US Federal Government.

Major lobbyists include the defence, tobacco and health-care industries. The National Rifle Association (NRA) benefits gun manufacturers and enthusiasts, but there is grave doubt as to their benefit to society as a whole, especially after the recent spate of mass shootings.

Climate change lobbyists have had some success in various countries benefiting some companies involved in alternative energy projects as well as the majority of stakeholders.

In the traditional sense, a think tank is a research institute that performs research and advocacy on topics such as social policy, political strategy, economics, military, technology and culture. Unfortunately, there is now a blurring of the lines between researchers and lobbyists and this new phenomenon is influencing radical politics (75audio).

## Pressure groups

Pressure groups can also play a part in changing a company's direction. Litigation has been used as a tool with over 1,000 climate change-related cases being filed in 25 countries.

The first major climate change lawsuit (128news) against a major oil company went to trial in the United States in October 2019. The New York Supreme Court considered the Attorney General's securities fraud claim against ExxonMobil for allegedly using two distinct sets of metrics to calculate the financial risks of climate change, one that was shared with investors and another that was used internally. The ExxonMobil lawsuit serves to highlight that the risk of climate related lawsuits is both real and potentially very costly. The Judge ruled against the Attorney General.

***Box 2.7*** outlines action against Royal Dutch Shell.

## Box 2.7: Royal Dutch Shell court summons (114news)

*"The Hague, April 5, 2019 - Today Friends of the Earth Netherlands will deliver **a court summons** to Shell to legally compel the company to cease its destruction of the climate, on behalf of more than 30,000 people from 70 countries. A 236 page complaint will be delivered to Shell's International Headquarters in the Hague this afternoon by Friends of the Earth Netherlands, ActionAid NL, Both ENDS, Fossielvrij NL, Greenpeace NL, Young Friends of the Earth NL, Waddenvereniging and a group of around 400 co-plaintiffs."*

*Roger Cox, leading Friends of the Earth's case against Shell said "If successful, the uniqueness of the case would be that Shell, as one of the largest multinational corporations in the world would be legally obligated to change its business operations. We also expect that this would have an effect on other fossil fuel companies, raising the pressure on them to change."*

**Box 2.8** shows the resistance against Japan's financing of coal projects.

## Box 2.8: G20 2019 Summit in Japan: Activists against coal

Jun 25, 2019 - Buy No Coal Japan (115website)

***Thousands worldwide mobilize against Japan's coal finance before G20 Summit***

*Thousands of activists from around the world – in Japan, Indonesia, India, the Philippines, Bangladesh, Pakistan, Australia, the US – have mobilized during the last week to protest Japan's continued financing of coal worldwide.*

*In addition to the in-person actions, over 80,000 people signed a joint petition calling on Japanese Prime Minister Shinzo Abe to end Japan's government-backed financing of coal projects. The petition was organized by SumOfUs, 350 Japan and Oil Change International. Activists have also protested the inaction of G20 governments more broadly on ending fossil fuel subsidies, to which they committed a decade ago.*

## Youths and social media

Social media has become a very powerful tool for change. In the early 2010s, the youth of the Arab world rose up against oppressive regimes and a low standard of living. It started in Tunisia by Mohamed Bouazizi's self-immolation in protest of police corruption and ill treatment. Social media played a significant role by facilitating communication and interaction among participants of the political protests.

Conversely, a major threat of social media is in the radicalization of the youth. Fake news promoted by radical websites and social media, is influencing vulnerable youths. Solutions could include participation in sport and other suitable leisure activities. Programs such as The Duke of Edinburgh's Award promote leadership skills and give motivation to succeed (see **Box 2.9)**.

> Box 2.9: Duke of Edinburgh's Award (116website)
>
> *The Duke of Edinburgh's Award, is a youth awards program which was founded in the United Kingdom in 1956 by Prince Philip, Duke of Edinburgh. It has since expanded to 144 nations.*
>
> *The International Award is a self-development program available to all young people worldwide equipping them with life skills to make a difference to themselves, their communities and the world.*
>
> *The personal program of activities includes helping others, pushing oneself physically, gaining skills and environmental/ conservation work.*

## Supra national groups

These are groups of nations and/or companies that have common interests.

### Davos World Economic Forum (76wiki)

The World Economic Forum (WEF) is an International Organization for Public-Private Cooperation. The Forum engages the foremost political, business and other leaders of society to shape global, regional and industry agendas.

## G20 Countries (77wiki)(78news)

The G20 is made up of a group of nations with the most advanced and emerging economies in the world. These countries make up about 85% of global GDP, over 75% of global trade, approximately two-thirds of the world's total population and about half of the earth's total land area.

## BRICS

BRICS is the acronym coined for an association of five major emerging national economies: Brazil, Russia, India, China and South Africa. Originally the first four were grouped as "BRIC", before the induction of South Africa in 2010. The original aim of BRIC was the establishment of an equitable, democratic and multipolar world order, but has since become a political organization, especially after the inclusion of South Africa.

## Trade groups

The World Trade Organization (WTO) is the only global organization that deals with the rules of trade between nations. At its heart are the WTO agreements, negotiated and signed by the bulk of the world's trading nations and ratified in their parliaments.

The European Union is a political and economic union of 28 member states that are located primarily in Europe.

ASEAN (Association of South East Asian Nations) is a regional intergovernmental organization comprising ten countries in South East Asia, which promotes intergovernmental cooperation and facilitates economic, political, security, military, educational and cultural integration among its members and other countries in Asia.

NAFTA (North American Free Trade Agreement) was negotiated between the United States, Canada and Mexico for the purpose of removing barriers to the exchange of goods and services among the three countries.

Mercosur is a South American trade bloc established by the Treaty of Asuncion in 1991 and Protocol of Ouro Preto in 1994. Its full members are Argentina, Uruguay, Paraguay and Brazil.

### International Civil Aviation Organization (ICAO) (117website)

The ICAO is a United Nations specialized agency, established in 1944 to manage the administration and governance of the Convention on International Civil Aviation (Chicago Convention).

The aviation industry has committed to reducing carbon emissions by 50% from their 2005 level by 2050. Aircraft currently contribute approximately 2.5% towards global warming and this figure is expected to rise to 3.5% by 2030.

Actions being taken to reduce the contribution of air travel to global warming include:

1. Stopping the growth of emissions by whatever means.

2. Using sustainable aircraft fuel in place of fossil fuel.

In 2020, the Covid-19 Pandemic dramatically reduced air travel emissions for several months.

### International Maritime Organization (IMO) (120website)

As a specialized agency of the United Nations, IMO is the global standard-setting authority for the safety, security and environmental performance of international shipping. Its main role is to create a regulatory framework for the shipping industry that is fair and effective, universally adopted and universally implemented. A specific regulation on air emissions is outlined in *Chapter 3 paragraph Disasters averted or postponed by a joint effort - Sulphur in air* .

## Individual influencers

Bill Gates invented Windows, which is now the de-facto operating system for personal computers. Sir Tim Berners Lee's invention of the world wide web has completely transformed the world.

The power of ideas should not be underestimated, but the right timing is essential. Many brilliant ideas have failed because of lack of customer acceptance at the time.

> Over the years there have been many attempts to make electric cars, with little success . It is only recently that the technology and infrastructure has been improved and developed to be acceptable to the mass market. The invention of the lithium battery and the forceful implementation of one individual, Elon Musk, has ensured that electric cars are making major inroads into the US and European markets.

I have listed examples of individuals that have influenced companies in various categories.

Entrepreneurs: Entrepreneurs are vital for the future development of business and need to be encouraged and supported. In Africa, entrepreneurs are being supported by organizations such as the African Entrepreneur Collective (130website) .

> When two surfers saw the appalling amount of plastic polluting the ocean around Bali, they were motivated to form a plastic collection, up-cycling and recycling operation. They are now a global for profit company funded by the sale of bracelets and other products made from recycled materials (129website).

Philanthropists: Individuals who have made their fortunes and are reinvesting in society, such as Bill Gates, founder of Microsoft (79website), and investors Howard Buffet (80website) (81book) and George Soros (82website).

Change agents: Individuals that promote change for the good of society and the earth. Muhammad Yunus (83website)(84website) founder of Grimeen micro banking, and ex US vice president Al Gore promoter of environmental issues are exemplary change agents.

Mavens: People that are turned to as experts in their field and can be the link in the transference of information. Tim Berners Lee (85website) inventor of the world wide web, and Jimmy Wales and Larry Sanger founders of Wikipedia (86website) are well known examples.

Activists: People who campaign to bring about political or social change. Examples are Nigerian Activist 'Ken' Saro-Wiwa (87news)(135news) (see **Box 2.10**) and Greta Thunberg (88news) a teenage environmental activist who has inspired teenagers around the world to protest against climate change and the inaction of those in power.

> Box 2.10: 'Ken' Saro-Wiwa
>
> *Saro-Wiwa was a member of the Ogoni people, an ethnic minority in Nigeria whose homeland in the Niger Delta is where crude oil has been extracted since the 1950s, causing extreme environmental damage.*
>
> *Initially as spokesperson, and then as president of the Movement for the Survival of the Ogoni People (MOSOP), Saro-Wiwa led a non violent campaign against environmental degradation of the land and waters of Ogoniland by the International Oil Companies (IOCs), including Shell. He was also an outspoken critic of the military dictatorship of Sani Abachi, who he considered derelict in enforcing environmental regulations on foreign petroleum companies.*
>
> *At the peak of his non-violent campaign, he was tried by a military tribunal for allegedly masterminding the gruesome murder of Ogoni chiefs. He was convicted and hanged in 1995. His execution provoked international outrage and resulted in Nigeria's suspension from the Commonwealth of Nations for over three years.*
>
> *Shell executives were accused of collaboration over the hanging of Saro Wiwa and subsequently paid $15.5 million to one group of activists' families, including the Saro-Wiwa estate. Shell denied any responsibility or wrongdoing.*

Investigative journalists: Those who uncover secrets that are in the public interest such as Clare Rewcastle-Brown (89website), founder of the Sarawak Report and Elliot Higgins (90movie) founder of Bellingcat.

Zealots: Those who have a driving passion to achieve their vision no matter what. There are many in Silicon Valley and a number of them 'fudge the figures' to get funding from venture capitalists. Examples are Elon Musk (91book) co-founder of the Tesla electric car and Elizabeth Holmes founder of the failed Theranos company which was meant to produce a mini lab for blood testing (92book) (93movie).

Political flies: Those who attempt to please politicians in power in exchange for favors. The Gupta Brothers (94wiki) attempted to effect 'State Capture' of the South African government by influencing the appointment of senior ministers in return for favors.

Corrupt politicians: Those who seek personal gain while in public office. Taib Mohamed (95website) current Governor of Sarawak, Malaysia has claimed a percentage of all logging contracts in Sarawak over the past 30 years, Najib Razak (96wiki) ex Prime Minister of Malaysia is regarded as the most significant money launderer ever, Jacob Zuma (97wiki) ex President of South Africa is estimated to have cost the South African economy approximately US$83 billion through his corrupt dealings and Dilma Rousseff (98news), ex President of Brazil was impeached after being involved in bribery and corruption, are examples.

Power hungry capitalists: Those who seek power and influence. Media mogul, Rupert Murdoch (99wiki), has influenced politics over decades, through the likes of the Sun newspaper in the UK and Fox News in the US.

Scam artists: Those who trick people out of their money. Examples are Nic Leason (100movie) (101website) who destroyed Barings Bank, Skilling and Lay (102website) of the Enron Scam, Jho Low (103news) who stole from the Malaysian Sovereign Fund (1MDB) and Bernie Madoff (104wiki) who set up a multi billion dollar Ponzi Scheme.

Whistleblowers: People who exposes confidential information or activities that are deemed illegal or unethical, within a private or public organization. John Doe (105wiki) revealed the Panama Papers, Xavier

Justo (106website) uncovered the theft of funds from the Malaysian Sovereign Fund (1MDB), and Edward Snowden (107movie) uncovered illegal CIA data mining of information on the population at large. **Box 2.11** outlines what whistleblowing entails.

### Box 2.11: WhistleBlowing

*Enables employees, customers, suppliers, managers or other stakeholders to raise concerns on a confidential basis in cases where conduct is deemed to be contrary to the company's or society's values. It may include:*

*1. Actions that may result in danger to the health and/or safety of people or damage to assets or environment*
*2. Unethical practices in accounting, internal accounting controls, financial reporting and auditing matters*
*3. Criminal offenses, including money laundering, fraud, bribery and corruption*
*4. Failure to comply with any legal obligation*
*5. Miscarriage of justice*
*6. Any conduct contrary to the* **ethical principles of the company** *embedded in company policies frameworks etc.*
*7. Any other legal or ethical concern and concealment of any of the above.*

*The reporting mechanism should be simple using telephone, email and/or web based and surface mail communication channels from anywhere in the world. It needs to be directed to an independent service provider who removes all indications as to the identity of the callers, before submission to designated persons in the company. The program must also be monitored by the Company Audit Committee.*

## Conclusion

Influencers can have a major positive or negative effect on a company and the economy. As primary stakeholders, the shareholders and owners want companies to be profitable, preferably over a long term. Long term means sustainability and sustainability means balancing what the company has taken from society and the environment with what it gives back.

Customers are becoming more vocal on social responsibility and environmental issues with social media providing a platform for communication.

Other influencers including governments, who set the legal framework in which the company can operate, as well as groups and individuals who are pushing companies to be more sustainable, can alter their success.

Banks are key to support investment and could play a huge role in rooting out corruption and dishonesty.

The ethical behaviour of influential individuals can have a major impact, by helping to determine the future of companies and countries.

THE DEATH OF BUSINESS AS USUAL

# CHAPTER 3: LEARNING FROM HISTORY

## Introduction

This chapter focuses on disasters caused mainly by corrupt practices, and how in some cases we have learnt to avert them. Learning from the past is key to helping us prevent and eliminate corruption which in many cases is the root cause of financial and economic disaster.

## Disasters averted or postponed by a joint effort

I list some examples of joint action taken by world governments to avert disasters. The first few examples relate to air pollution, followed by pollution of the oceans, forests and, lastly, the damage agricultural and industrial practices wreak.

### Ozone depletion (2wiki)

When a hole in the earth's ozone layer was detected in 1985, all nations came together to ban CFCs which were used as refrigerants and fire extinguishing agents. Industry quickly found alternatives to keep fridges working.

The Montreal Protocol on Substances that Deplete the Ozone Layer is an international treaty designed to protect the ozone layer by phasing out the production of numerous substances that are responsible for ozone depletion. It was agreed on 26 August 1987 and, due to its widespread adoption and implementation, it has been hailed as an example of exceptional international co-operation.

> "perhaps the single most successful international agreement to date has been the Montreal Protocol". UN Secretary General Kofi Annan

The 2019 Ozone hole is the smallest on record since its discovery (96website).

## Lead in the air (3wiki)

Tetra ethyl lead (TEL), an organolead compound, was extensively used as a gasoline additive beginning in the 1920s, wherein it served as an effective antiknock agent that allowed engine compression to be raised substantially. Because of concerns over air and soil lead levels, and the accumulative neuro-toxicity of lead, most industrialized countries had phased out TEL from gasoline by the early 2000s.

## Green House Gas (GHG) emissions

To address the issues of global warming, the United Nations Framework Convention on Climate Change (UNFCCC) (5website) was adopted in 1992, with the objective of limiting the concentration of Green House Gases (GHGs) in the atmosphere. The Kyoto Protocol came into force in February 2005 as a supplement to the Convention to set limits on the maximum amount of emissions by countries. The Kyoto Protocol committed 41 countries to reduce their GHG emissions by at least 5% below their 1990 baseline emission within the commitment period of 2008 to 2012. As per the Kyoto Protocol, least developed and developing countries are not bound to reduce GHG emissions.

GHG emissions are required to be reported in $CO_2$ equivalent (CO2e). This is measured in terms of the Global Warming Potential (GWP), which is the heat absorbed by any greenhouse gas in the atmosphere, as a multiple of the heat that would be absorbed by the same mass of carbon dioxide. The GWPs of the various GHGs are shown in **Table 3.1**.

Table 3.1 Global Warming Potentials for GHGs (4website)

| S | Greenhouse Gas (GHG) | Chemical Formula | Global Warming Potential (GWP) for 100 Year Time Horizon |
|---|---|---|---|
| 1 | Carbon dioxide | $CO_2$ | 1 |
| 2 | Methane | $CH_4$ | 21 |
| 3 | Nitrous Oxide | $N_2O$ | 310 |

Greenhouse gases can be traded to reduce a country's or a company's emissions. Trading is done in units of Certified Emission Reductions (CERs). One CER = 1 metric ton of $CO_2$ equivalent.

Keeping $CO_2$ out of the atmosphere via carbon capture and storage (CCS) is viewed by many scientists as necessary to limit global warming to the lower target of 1.5 degrees Celsius, which was set in the 2015 Paris Agreement to tackle climate change. Scores of entrepreneurs (6news) have been wracking their brains to move beyond the most common way to reuse waste $CO_2$ - pumping it into the ground to extract oil - and deploy it instead to make more planet-friendly materials.

## Sulphur in the air (9website)

In a landmark decision for both the environment and human health, 1 January 2020 was set as the implementation date for a significant reduction in the sulphur content of the fuel oil used by ships.

The decision to implement a global sulphur limit of 0.50% m/m (mass/mass) from 2020 was taken by the International Maritime Organization (IMO), the regulatory authority for international shipping, in 2016.

## Plastic in the sea

It is now recognized that plastic pollution of the sea is untenable. Many organizations are now focused on removing plastic from the sea, while others are committed to preventing non-biodegradable plastics from entering the packaging market. A few examples follow.

The International Union for Conservation of Nature (IUCN)(10website) is a membership union uniquely composed of both government and civil society organizations.

IUCN gives the following statistics:

- Over 300 million tons of plastic are produced every year for use in a wide variety of applications.

- At least 8 million tons of plastic end up in our oceans every year, and make up 80% of all marine debris from surface waters to deep-sea sediments.
- Marine species ingest or are entangled by plastic debris, which causes severe injuries and deaths.
- Plastic pollution threatens food safety and quality, human health, coastal tourism, and contributes to climate change.
- There is an urgent need to explore the use of existing legally binding international agreements to address marine plastic pollution.
- Recycling and reuse of plastic products, and support for research and innovation to develop new products to replace single-use plastics are also necessary to prevent and reduce plastic pollution.

The Alliance For Plastic Waste (11website) is mentioned in **Chapter 2**. Plastics recycling is on the increase and is discussed further in **Chapter 5**. Efforts have got a long way to go to turn the tide of plastic pollution in the sea.

## Whale population

The International Whaling Commission (IWC) (12website) was set up under the International Convention for the Regulation of Whaling which was signed in Washington DC in December 1946. In 1982 the IWC decided that there should be a pause in commercial whaling on all whale species and populations from the 1985/1986 season onwards. This pause is often referred to as the commercial whaling moratorium, and it remains in place today. However, Iceland and Norway continue to undertake commercial whaling, and Japan (13news) left the IWC in June 2019 as it chose not to comply with the requirements of the IWC. It continues to catch whales in the Antarctic.

## Forests

After the worst Borneo fires in 2018, the governments of Singapore,

Malaysia and Indonesia agreed on approaches to control the fires, but did not stop the cutting and burning of the tropical forests. In 2019 the fires were even worse.

Other efforts include work done by Willie Smits in Indonesia (14wiki) and the governments and people of Ethiopia, China, Thailand, Spain, Norway, Israel. New Zealand, Costa Rica, Iran, Kenya, Philippines, Nigeria, France, Malaysia, Peru, Cuba, USA and Mexico. Some news items follow.

- Ethiopia says it planted over 350 million trees in a day, a record (15news).
- Extensive reforestation in China makes Earth greener (16news).
- Thailand: New laws chased to boost afforestation (17news).

Reforestation has got a long way to go before we see the fruits of our efforts.

## Agriculture

The use of certain pesticides in agriculture has been seriously detrimental to the environment over the long term (90book) (7video). The use of DDT (dichloro diphenyl trichloroethane) was banned as a pesticide worldwide under the Stockholm Convention in 2001 after it was discovered to be dangerous to wildlife and the environment, but it has taken over 25 years to reduce its toxicity to acceptable levels.

## Chemicals in the ground

Many harmful and toxic chemicals have been dumped into the ground and waterways. One of the more serious pollutants are polychlorinated biphenyls (PCBs) which were used in hundreds of industrial and commercial applications, including electrical insulating fluids in transformers and plasticizers in paints, plastics and rubber products. They were banned in the US in the late 1970s and as a result PCB levels in the environment and humans are much lower, but still pose a danger to our health (91news).

## Lessons to be learned from serious incidents

The focus on profit alone has led some major companies to take large risks resulting in disaster. Pressure from shareholders and other stakeholders must be put on company management to prevent excesses.

Primary areas of pollution have been the sea, rain forests and the atmosphere. The primary pollutants have been from oil, chemicals and radioactivity, while atmospheric and water pollution have been the main culprits affecting human health. The most notable disasters can be categorized as follows:

- Oil Exploration
- Chemical Manufacturing
- Nuclear Energy
- Mining
- Logging
- Water and Food Resources
- Financial

Examples are listed as follows.

### Oil exploration

Oil spillage has a major impact on the ecosystem into which it is released and may constitute ecocide. Ecocide literally means killing the environment - causing extensive damage or destroying ecosystems or harming the well-being of a species. It has not yet been accepted as an international crime by the United Nations.

The worst culprits in polluting the environment from oil exploration include the following.

Saddam Hussain Dictator of Iraq: Kuwait oil fires (18news)

As the 1991 Persian Gulf War drew to a close, Saddam sent men to blow up Kuwaiti oil wells. Approximately 600 wells were set ablaze, and the fires burned for seven months. The Gulf was awash in poisonous smoke, soot and ash. Black rain fell and lakes of oil were created. A twilight descended on the Gulf that year and dates did not even ripen that summer. The lakes of oil are still there to this day.

Exxon Mobil: *Exxon Valdez* Alaska oil spill (19news)

On 24th of March 1989, the *Exxon Valdez* oil tanker ran aground in the pristine waters of Alaska's Prince William Sound and released 10.8 million gallons of oil. It would eventually spread almost 500 miles from the original crash site and stain thousands of miles of coastline. Hundreds of thousands of birds, fish, seals, otters and other animals perished as a result, despite the mobilization of more than 11,000 people and 1,000 boats as part of the cleanup.

**The *Exxon Valdez* oil leak was considered to be the largest man-made environmental disaster in U.S. history at the time.**

British Petroleum: Deepwater Horizon Gulf of Mexico oil spill (20movie) (21book)

In April 2010, the BP Deepwater Horizon semi submersible drilling platform sank after exploding and burning. Eleven lives were lost. Two days after the initial explosions, the rig sank. On the ocean floor, the wellhead erupted spewing an estimated 200 million gallons of oil in the next ten weeks - the equivalent of 20 Exxon Valdez spills. **This is now regarded as the biggest ever oil spill to date.**

President Obama subsequently abolished the Interior Department's Mineral Management Service (MMS). He split it into 3 agencies where one issued leases, the second collected royalties and the third supervised offshore drilling and production.

Most of the spill is still lurking on the bottom of the Gulf of Mexico to this day.

BP has been ordered to pay $5.5 billion to settle civil damages claims made by the U.S. as a result of the Deepwater Horizon oil spill. The amount will be paid over the course of 16 years.

**Obviously people in power did not learn from the Exxon Valdez disaster.**

Texaco: Ecuador rainforest pollution (22website)(23movie)

Texaco, an affiliate of Chevron, is responsible for creating toxic contamination 30 times larger than the Exxon Valdez. **Probably the largest oil-related environmental catastrophe in the world exists quietly in the Amazon rainforest, threatening to wipe out five indigenous groups largely out of sight of the world's media.** *Box 3.1* outlines the saga.

> Box 3.1: Chevron Ecuador 'Oil Dumping' case (edited) (24news)
>
> *BBC 8 Sept 2018.*
>
> ***Chevron wins Ecuador rainforest 'oil dumping' case***
> ***An international tribunal in The Hague has ruled in favour of the US oil company, Chevron, in an environmental dispute with the government of Ecuador.***
>
> *Chevron had been ordered to pay $9.5bn compensation to thousands of residents in Ecuador's Amazon region.*
> *The residents accused the company of dumping toxic waste in local lakes and rivers of the Lago Agrio region for decades.*
> *The international tribunal said that the 2011 Ecuador Supreme Court ruling had been obtained through fraud, bribery and corruption.*
> *The oil giant now stands to be awarded hundreds of millions of dollars in costs by the tribunal.*
> *Chevron maintained that it never owned any assets in Ecuador. The alleged environmental damage was done by Texaco between 1964 and 1992. Texaco was later acquired by Chevron…..*
>
> *Some 30,000 local residents, including five different Amazonian tribes, began the lawsuit against Texaco in 1993. The plaintiffs say that the oil company knowingly dumped 18 billion gallons of toxic*

*waste water and spilled 17million gallons of crude oil into the rainforest during its operations in north-east Ecuador. ...the affected area covers 4,400 sq km on the border with Colombia. Local residents believe the pollution has led to health problems such as cancer and birth defects.*

*After the latest ruling in the Hague, a lawyer for the indigenous communities criticized the Ecuadorean government for agreeing to take the case to an arbitration court."That is playing Chevron's game and leaving the crime unpunished forever," said Pablo Fajardo. He said he was considering all legal avenues.*

Royal Dutch Shell and other International Oil Companies: Niger Delta pollution (25wiki)

Oil spills are a common event in Nigeria. Half of all spills occur due to pipeline and tanker accidents. Other causes include sabotage (28%) and oil production operations (21%), with 1% of the spills being accounted for by inadequate or non-functional production equipment. A UNDP report states that there have been a total of 6 817 oil spills between 1976 and 2001, which account for a loss of three million barrels of oil, of which more than 70% has **not** been recovered. 69% of these spills occurred off-shore, 25% were in swamps and 6% on land.

Ken Saro-Wiwa attempted to stop the destruction of the Niger Delta environment and was hanged by the then dictator of Nigeria (see **Box 2.10**). Shell was implicted in his hanging.

Shell has recently been issued a court summons to cease destruction of the climate (see **Box 2.7**).

## Chemical manufacturing

I have outlined a few of the worst chemical disasters.

Union Carbide: Bhopal India disaster (26news)

On 2nd December 1984, an accident at a Union Carbide pesticide plant in Bhopal, India, resulted in 45 tons of poisonous methyl isocyanate (MIC) escaping into the atmosphere. Thousands died within hours with more following in subsequent months, approximately 15,000 in all. About half a million people were affected in some way and a shockingly high number of children in the area have been born with birth defects.

In 1989, Union Carbide paid out half a billion dollars to victims, an amount the afflicted say is not nearly enough to deal with the decades-long consequences. **Bhopal remains the worst industrial disaster in history.**

ICMESA/ Givaudan/ Hoffmann-La Roche Italy: Seveso Dioxin Cloud (27news)

On 10th July 1976, an explosion at an Italian chemical plant released a cloud of dioxin that settled on the town of Seveso, north of Milan. At first animals began to die and just 4 days after the explosion people began feeling ill. Only weeks later was the town evacuated.

The disaster at Seveso resulted in industrial safety regulations being passed in the European Community in 1982 called the Seveso Directive, which imposed much harsher industrial regulations. The Seveso Directive was updated in 1996, 2008 and 2012 and is currently referred to as the Seveso III Directive (or COMAH Regulations in the United Kingdom).

## Nuclear power

Nuclear power is touted to be very safe, but when a disaster happens, it affects mankind for a very long time. The worst of these are listed.

Chernobyl USSR State Power Company Ukraine (28movie)

On 26th April 1986, reactor number four at the Chernobyl nuclear power plant exploded. More explosions ensued, and the fires that resulted sent radioactive fallout into the atmosphere, four hundred times greater than that of the atomic bombing of Hiroshima.

A New Safe Confinement (NSC or New Shelter) (8wiki) was built to confine the remains of the number 4 reactor unit. It was completed in 2019.

**The Chernobyl disaster is considered to be the worst nuclear power plant disaster in history.**

Tokaimura Nuclear Plant Japanese Research Facility (29video)

On 30th September 1999, Japan's worst nuclear accident at the time happened in a facility northeast of Tokyo. Three workers at a uranium-processing plant in Tokaimura improperly mixed a uranium solution. Two died, and hundreds were exposed to various levels of radiation.

Tokyo Electric Power Company (TEPCO) Japan: Fukushima

This was the biggest nuclear meltdown since Chernobyl (30news). The March 2011 incident is described in ***Box 3.2***.

### Box 3.2: Fukushima Nuclear Reactor meltdown chain of events (31wiki)

*The 9.0 $M_W$ Tōhoku earthquake occurred at 14:46 on Friday, 11th March 2011. The earthquake triggered a 13 to 15 meter high tsunami that arrived approximately 50 minutes later. The waves flowed over the plant's 5.7 meter seawall flooding the basements of the power plant's turbine buildings and disabling the emergency diesel generators. The insufficient cooling eventually led to meltdowns in Reactors 1, 2, and 3.*

*Reactor 4 was not operating when the earthquake struck. All fuel rods from Unit 4 had been transferred to the spent fuel pool on an upper floor of the reactor building prior to the tsunami. On 15th March, an explosion damaged the fourth floor rooftop area of Unit 4, creating two large holes in a wall of the outer building.*

*The Fukushima reactors were not designed for such a large tsunami, nor had the reactors been modified when concerns were raised in Japan and by the International Atomic Energy Agency (IAEA). An in-house TEPCO report in 2000 recommended safety measures against seawater flooding, based on*

> the potential of a 15 metre tsunami. TEPCO leadership said the study's technological validity "could not be verified." A 2008 in-house study identified an immediate need to better protect the facility from flooding by seawater. This study mentioned the possibility of tsunami-waves up to 10.2 meters. Headquarters officials insisted that such a risk was unrealistic and did not take the prediction seriously.

Today the reactors are decommissioned but radioactive water has accumulated and may still have to be released into the Pacific.

The incident was rated 7 on the International Nuclear Event Scale (INES). This scale runs from 0, indicating an abnormal situation with no safety consequences, to 7, indicating an accident causing widespread contamination with serious health and environmental effects. Prior to Fukushima, the Chernobyl disaster was the only level 7 event on record, while the Three Mile Island accident in the USA was rated as level 5.

**People in power obviously did not learn from Chernobyl and Three Mile Island.**

## Mining

In many countries a mining concession requires that the natural environment be reinstated after completion of mining. In the Tar Sands of Canada (93news), oil sands producers must, by law, reclaim the land when they are finished with their operations. Yet only a small fraction of the mined land has been reclaimed so far. This is the world's most destructive oil operation, and it's growing.

> "Large enough to be seen from space, tailings ponds in Alberta's oil sands region are some of the biggest human-made structures on Earth. They contain a toxic slurry of heavy metals and hydrocarbons from the bitumen separation process." (93news)

The long term effect on the health of miners is also detrimental to the communities. An example follows.

## Asbestos Mining: class action (32website)(33website)

Cape Industries was taken to the High Court in the UK by the residents of the mining town in the Northern Cape of South Africa. The saga is described in *Box 3.3*.

### Box 3.3: Cape Asbestos Class Action (edited)

*In 1997, a group of five South Africans suffering from asbestos-related disease (ARD) brought suit against Cape PLC in the English High Court seeking compensation for their injuries from Cape's asbestos mining and milling activity in South Africa. In 1999, another 2000 claims were commenced against Cape in England for ARD based on Cape's activity in South Africa. By 2001 there were approximately 7500 claimants. In 2001, Cape agreed to a £21 million out-of-court settlement with the plaintiffs, but the company encountered financial problems in August 2002 and did not meet the agreed settlement terms. Therefore, the litigation recommenced in September 2002, and Gencor Ltd. was joined as a defendant in the case. Gencor is a South African company which took over some of Cape's South African asbestos operations when Cape left the country in 1979.*

*In 2003, the plaintiffs, Cape and Gencor reached a settlement agreement.*

Asbestos is listed as a category of controlled waste under Annex I of the Basel Convention on the Control of Transboundary Movements of Hazardous Wastes and their Disposal [1992] (34wiki). Specifically, any waste streams having asbestos (dust and fibers) as constituents are controlled. In general, Parties to the Convention are required to prohibit the export of hazardous wastes.

## Logging

Due to corruption in many countries, logging of tropical rain forests is out of control.

> *"50% of Borneo's rainforest has been lost in the past 50 years due to logging"* (35book)

Burning takes place after the logging is done to clear all remaining vegetation before planting the palm oil trees. The governments of

Singapore, Malaysia and Indonesia put out a health warning at the height of the burning of forests in 2018 and 2019 as the whole region was covered in smoke (36wiki).

Evidence of the extent of damage in Borneo (38website) has been documented by Center for International Forestry (CIFOR). The book promoted in **Box 3.4** outlines the situation in Sarawak Borneo.

> ### Box 3.4: 'Money Logging' by Lukas Straumann (37book)
>
> *Money Logging investigates what former British Prime Minister Gordon Brown has called 'probably the biggest environmental crime of our times'—the massive destruction of the Borneo rainforest by Malaysian loggers. Historian and campaigner Lukas Straumann goes in search not only of the lost forests and the people who used to call them home, but also the network of criminals who have earned billions through illegal timber sales and corruption.*
>
> *Straumann singles out Abdul Taib Mahmud, current governor of the Malaysian state of Sarawak, as the kingpin of this Asian timber mafia, while he shows that Taib's family—with the complicity of global financial institutions—have profited to the tune of 15 billion US dollars. Money Logging is a story of a people who have lost their ancient paradise to a wasteland of oil palm plantations, pollution, and corruption—and how they hope to take it back.*
>
> *"If there were a list of the world's leading environmental criminals, Sarawak's government head Taib Mahmud would surely be in the top ten. No other person bears a larger individual responsibility for the destruction of the tropical rain forests on Borneo, one of the Earth's most prolific lebensräume."*
>
> *"During Taib's 30 years in office as Chief Minister of the Malaysian state Sarawak, bulldozers belonging to the timber and plantation conglomerates have **reduced** the former primeval forests **to** an area of about **5% of their original size**." (40website)*

In Sarawak, the "Big Six" timber firms (39news) have a combined concession areas of 3,724,675 hectares of forest land, with Samling

Group of Companies holding the biggest concessions with 1,288,389ha, followed by RH Group of Companies (1,001,877ha), Shin Yang Group of Companies (500,904ha), Ta Ann Group of Companies (433,003ha), WTK Group of Companies (357,017ha) and KTS Group of Companies (144,485ha).

> "the Norwegian Government Pension Fund, has sold all its 16 million shares of Malaysian timber giant Samling Global.......found the corporation responsible for systematic illegal logging in .... Sarawak on Borneo." (43website)

In Liberia, (41website) a quarter of its total landmass has been granted to logging companies in just two years, following an explosion in the use of secretive and often illegal logging permits. Subsequently, Taib's cousin, the logging tycoon and key political figure Hamed Sepawi, has been booted out of Liberia by a Presidential decree (42website) to halt corrupt and illegal logging.

Meanwhile Borneo Indonesia has now initiated certification for export of it's hard wood. (See **Box 1.16**)

The Amazon has been seriously affected by actions of the new government of Brazil which has permitted the recommencement of logging and burning in the Amazon causing major pollution in the region (91wiki).

## Water and food resources

I outline a few of the worst examples of contamination of the earth's water & food supply and the health of its population.

### Love Canal Chemical Companies: USA (44news)

In 1978, Love Canal, located near Niagara Falls in upstate New York, was a pleasant little working-class enclave consisting of hundreds of houses and a school. It just happened to sit on top of 21,000 tons of toxic industrial waste that had been buried underground in the '40s and '50s by a local company. Over the years, the waste began to bubble up and by 1978, the problem resulted in hundreds of families selling their houses to the US Federal Government. The disaster led to the formation in 1980 of the Superfund program, which helps pay for the cleanup of toxic sites.

## The Aral Sea USSR Government (94news)

Situated between Uzbekistan and Kazakhstan, the Aral was once the fourth largest lake on earth, as big as Ireland. Since the 1960s, Soviet irrigation projects diverted several of its source waterways, resulting in it shrinking by 90%. What was once a vibrant, fish-stocked lake is now a massive desert that produces salt and sandstorms that kill plant life and have negative effects on human and animal health for hundreds of miles around. Uzbekistan and Kazakhstan are now taking joint action to try to refill the sea.

## Pacific Gas and Electric Company (PG&EC) USA: Hinkley groundwater contamination (45wiki)(46movie)

From 1952 to 1966, PG&EC dumped about 370 million gallons of chromium tainted waste-water into unlined waste-water spreading ponds around the town of Hinkley, north of Los Angeles.

In 1993, legal clerk Erin Brockovich began an investigation into the health impacts of the contamination. A class-action lawsuit was settled in 1996 for $333 million, the largest settlement of a direct action lawsuit in U.S. history. Hinkley is now a ghost town.

## Du Pont Ohio River contamination: class action (47movie)

Over many years Du Pont's waste water, flowing into the Ohio river, has been found to be the cause of a high incidence of cancer in most of the population downstream of the plant. A class action was won against Du Pont in 2018.

This is apparently the tip of the iceberg with respect to contamination by perfluorinated chemicals, the ingredient for our 'Teflon®' non-stick pans and 'Gor-Tex®' raingear.

## City of Flint, Michigan: lead in drinking water (48news)(49news)

The water contamination in Flint is outlined in **Box 3.5**.

> ### Box 3.5: Flint water contamination
>
> *The city announced that a new pipeline would be built to deliver water from Lake Huron so as to reduce the water shortfall. In 2014, while the pipeline was under construction, the city turned to the Flint River as a water source. Soon after the switch, residents reported changes to the water's color, smell and taste.*
>
> *Tests in 2015 by the Environmental Protection Agency (EPA) and Virginia Tech indicated dangerous levels of lead in the water at residents' homes.*
>
> *More than a dozen lawsuits were filed against Michigan and the City of Flint, various state and city officials, employees involved in the decision to switch the source of the drinking water and those responsible for monitoring water quality. The range of remedies sought included monetary compensation for lead poisoning and refunds for water bills.*

In 2019 it was reported that the Flint water crisis is nowhere near over.

## Chisso Corporation Japan: Minamata Disease (50movie)

Japan's most serious industrial pollution-induced illness is outlined in **Box 3.6**.

> ### Box 3.6: Minamata disease
>
> *For years, residents of Minamata, a town located on Kyushu (Japan's most southwesterly island), had observed odd behavior among animals. People referred to the behavior as "cat dancing disease."*
>
> *In 1956, the first human patient was identified. Symptoms included convulsions, slurred speech, loss of motor functions and uncontrollable limb movements. Three years later, an investigation concluded that the affliction was a result of industrial poisoning of Minamata Bay by the Chisso Corporation. As a result of waste water pollution by the plastic*

*manufacturer, large amounts of mercury and other heavy metals found their way into the fish and shellfish that comprised a large part of the local diet. Thousands of residents have slowly suffered over the decades and died from the disease.*

Over sixty years since the first case was discovered the issue of relief for victims of the disease remains unresolved (93news).

## Fracking

With the removal of restrictions on 'fracking', the US has again become a net exporter of oil & gas. 'Fracking' has resulted in contamination of drinking water in large parts of the US (1movie).

## **Financial disasters**

I list a few major financial disasters which have destroyed the economic well being of communities and countries.

### 1997 Asian Flu (54wiki)

The Asian financial crisis was a period of financial crisis that gripped much of South East Asia beginning in July 1997 and raised fears of a worldwide economic meltdown due to financial contagion (see also **Box 1.6**).

### 2004 dot-com crash (55wiki)

The dot-com bubble was a historic period of excessive speculation mainly in the United States that ended in 2004, a period of massive growth in the use, and adoption of the Internet.

### Enron, Worldcom etc. 2002 (56wiki)

Enron was formed in 1985 by Kenneth Lay after merging Houston Natural Gas and InterNorth. When Jeffrey Skilling was hired, he developed a staff of executives that, by the use of accounting loopholes, special purpose entities, and poor financial reporting, were able to hide billions of dollars in debt from failed deals and projects.

The Enron scandal precipitated the Sarbanes-Oxley Act (57news), a Federal law that initiated a comprehensive reform of business financial practices. This act set new standards for public accounting firms, corporate management, and corporate boards of directors.

## 2007-8 Financial crisis (59book)

The financial crisis of 2007–2008 is considered by many economists to have been the most serious since the Great Depression of the 1930s.

As a result, all Icelandic banks went insolvent after Wikileaks exposed their financial situations and a number of their directors were sent to jail as a result. In the US, two major banks, Lehman Brothers and Washington Mutual, were permitted to go insolvent, while the Federal government bailed out others. No US bank officials were jailed. In the UK, banks were bailed out by the government and the Financial Services Authority (FSA) was restructured.

> "'Deficient' and 'inadequate' FSA failed to stop HBOS collapse. The Financial Services Authority was 'deficient' and 'inadequate' in its regulation of HBOS, which contributed to the collapse of the bank in 2008, according to a new report. ... The FSA was the regulator at the time. It split into the FCA and PRA in 2013."

The US Financial Crisis Inquiry Commission (58website) reported its findings in January 2011. It concluded that the crisis was avoidable and was caused by:

1. Widespread failures in financial regulation, including the Federal Reserve's failure to stem the tide of toxic mortgages

2. Dramatic breakdowns in corporate governance including too many financial firms acting recklessly and taking on too much risk

3. An explosive mix of excessive borrowing and risk by households and Wall Street that put the financial system on a collision course with crisis

4. Key policy makers ill prepared for the crisis, lacking a full

understanding of the financial system they oversaw and

5.Systemic breaches in accountability and ethics at all levels.

## The elimination of corruption by joint effort

The effect of corruption on government and citizens can be devastating (60video)(61tedtalk).

> "Corruption chips away at democracy to produce a vicious cycle, where corruption undermines democratic institutions and, in turn, weak institutions are less able to control corruption" (62website)

Further to **Chapter 1 Paragraph "3.Ethical Leadership and eradication of corruption"**, various laws have been enacted, especially by the USA and UK. After the Siemens scandal Germany tightened its anti-corruption laws.

The OECD Anti-Bribery Convention (63website) has gone a long way to reducing international corruption. It requires adherents to criminalize acts of offering or giving bribes, but not for soliciting or receiving bribes.

The World Bank is a vital source of financial and technical assistance to developing countries around the world. **Box 3.7** outlines the World Bank's first sanctions report in its fight against corruption.

> Box 3.7: World Bank Sanctions Report (92website)
>
> PRESS RELEASE OCTOBER 3, 2018
>
> 'The World Bank Group today affirmed its commitment to fighting corruption and safeguarding donor resources, showing in its first joint Sanctions System Annual Report the institution's significant efforts to investigate and adjudicate allegations of fraud and corruption in Bank Group projects. The Bank Group debarred 78 firms and individuals during fiscal year 2018, ...'

## Lessons to be learned from major corruption

*"the trade in weapons accounts for around 40 % of all corruption in all world trade"* (118movie)

Sadly, there are so many examples of gross corruption around the world that it is very difficult to list and describe those that we are now aware of. So I'll mention just a few.

This is a list of the ten biggest corruption related accounting scandals (99website) in recent times

1. Waste Management Inc 1998
2. Enron 2001
3. Worldcom 2002
4. Tyco 2002
5. HealthSouth 2003
6. FreddieMac 2003
7. AIG 2005
8. Lehman Brothers 2008
9. Bernie Madoff 2008
10. Satyam 2009

Except for Satyam, these are all American companies, possibly since there is more transparency within the American system, but in my opinion, lax governance by US companies is a major contributing factor which enables corruption (see also **Box 1.4**). A few American citizens, including Bernie Madoff and former Enron CEO Jeff Skilling (100wiki), were sent to prison and financial services companies Lehman Brothers and Washington Mutual (101news) (not listed) were bankrupted in the 2008 financial crisis. Corrupt auditing practice were another major factor. India's Ministry of Corporate Affairs banned auditing companies PricewaterhouseCoopers and Deloitte in the wake of the Satyam scandal. Further to PricewaterhouseCoopers ban in India, a number of the other big auditing companies have also got into hot water.

Arthur Andersen (AA) (102news) once counted itself as one of the world's "big five" accounting firms, until its criminal handling of the energy company Enron led to its downfall. In spite of this, former partners bought the rights to the Arthur Andersen name and started operating again 12 years later.

By a strange twist of fate, Ernst and Young fortuitously escaped being caught up in the 1MDB scandal when it was fired by Najib Razak, the then Prime Minister and Finance Minister of Malaysia.

KPMG International (103website) (104book) is a Swiss cooperative and the coordinating entity for a global network of independent firms. KPMG Malaysia were appointed after E&Y and were, unfortunately, immersed in the Malaysia 1MDB scandal, although they too were subsequently dismissed by Najib Razak.

Deloitte (105news) took over from KPMG in Malaysia and was fined for failing to uncover a massive corruption scandal related to Malaysia's Sovereign Wealth Fund 1MDB.

State Sovereign Wealth Funds generally are well managed, but can be prone to exploitation.

> "A sovereign wealth fund (or sovereign investment fund) is a state-owned fund that invests in real and financial assets such as stocks, bonds, real estate, precious metals, or in alternative investments such as private equity or hedge funds."

The Sovereign Wealth Funds of Norway and Qatar are examples of how successfully countries can manage their funds, but a number of countries have mismanaged their funds as we can see in these two examples.

Malaysia's Sovereign Wealth Fund, 1MDB, is the subject of corruption and money-laundering investigations in at least six countries. About US$4.5 billion is believed to have been siphoned from the fund, with about US$700 million of that diverted into ex Prime Minister Najib Razak's personal bank accounts. Banks (104book) implicated include

Goldman Sachs, Coutts, BSI, Ambank (majority owned by ANZ), J.P.Morgan, Standard Chartered, Falcon, Rothschild, Julius Baer and Deutsche Bank. Deutsche Bank has also been involved in a number of Trump-related scandals (106movie), besides being fined for conspiring to manipulate interest rates. This fine has been equated to an average American getting a $10 parking fine (107news).

Libya's oil wealth should have been invested for the future through it's Sovereign Wealth Fund (108news). However, the fund managers have mismanaged the money through nepotism, incompetence and wild gambling.

I have selected a significant few corruption scandals which are described in categories.

- Slush Funds
- Money Laundering
- State Capture
- Whistleblowing
- Bribes and Political Donations
- Minerals Exploitation
- Dictators and Corrupt Politicians

## Slush funds

### French Companies: Thales/ Thomson-CSF/ THEC/ Naval Group/ DCNS

Former executives of THEC, a subsidiary of Thales, alleged that Thales has a centralized **slush fund** that it used to bribe officials (66wiki).

In South Africa, Schabir Shaik, the financial advisor to the Deputy President of the African National Congress (ANC) party Jacob Zuma, was found guilty of organizing a bribe on behalf of Thomson-CSF. Shaik was sent to prison but was released on parole after just a few years (64wiki).

In Malaysia, DCNS is alleged to have paid more than USD 134 million in kickbacks to a shell company linked to Abdul Razak Baginda, a close associate of the ex Malaysian Prime Minister, Najib Razak (65news).

In Taiwan in the early 90s, French state owned Elf Aquitaine was involved (along with other companies and countries) in selling frigates to Taiwan. In June 2011 Thales Group and the French Government were ordered to pay 630 million euros (almost a billion US dollars) in fines after the courts heard that bribes had been paid to the Taiwanese government to win this large naval contract. To this day, this is the largest corruption case in French history (66wiki).

### German Company: Siemens (67news)

For over a decade, Siemens paid bribes to government officials and civil servants around the world, amounting to approximately US$1.4 billion. While corrupt decision makers profited, citizens in the affected countries paid the costs of overpriced necessities such as roads and power plants.

> "... unprecedented in scale and geographic reach. The corruption involved more than $1.4 billion in bribes to government officials in Asia, Africa, Europe, the Middle East and the Americas."
>
> Security Exchange Commission (SEC) director, Linda Thomsen

Action was finally taken against Siemens in a number of countries including the US, Germany, Italy and Lichtenstein. Following the US and German prosecutions, Siemens paid more than $1.6 billion in fines, penalties and disgorgement of profits, including $800 million to US authorities. This was the largest monetary sanction ever imposed in a case under America's Foreign Corruption Practice Act since it was passed in 1977.

## Money laundering

### Malaysia: Najib Razak (68book) and 1MDB

It is estimated that more than US$4 billion was embezzled in what is one of the world's biggest known corruption schemes.

In 2009, the government of Malaysia set up a Sovereign Wealth Fund, 1Malaysia Development Berhad (1MDB). Chaired by the former prime minister, Najib Razak (also the finance minister at the time), the fund was originally meant to boost the country's economy through strategic investments. But instead, it seems to have boosted the bank accounts of a few individuals. Jho Low, one of the king pins, is on the run, while another key player, Khadem al-Qubaisi, was arrested in the United Arab Emirates in 2016. In 2019, he was sentenced to 15 years in prison for corruption and money laundering. A few bankers have been arrested in Singapore and Najib Razak is currently appearing in court. Legal action is ongoing in Switzerland, USA and Singapore.

### Russian Laundromat (69news)

The Russian Laundromat was a massive money laundering scheme that siphoned off between US$20 and 80 billion in fraudulent funds. Formal investigations are currently underway in several countries and the banks involved – Moldindconbank, Danske Bank, Deutsche Bank and HSBC – are in hot water for failing to comply with anti-money laundering rules.

### Venezuela: Currency exchange scam (70news)

Less than 20 years ago, Venezuela was South America's richest country. Today it's facing one of its worst political and humanitarian crises, and corruption has a key role in it. The plundering of the National Oil Company, PDVSA, is an example of the widespread corruption at the highest levels of government.

A number of prominent members of the Venezuelan elite were involved in a scheme where they played the country's currency system to gain money illegally, totaling $1.2 billion in fraudulent sums. Among them, according to US investigators, are Raúl Gorrín, a Venezuelan billionaire who owns the news network Globovision, and three stepsons of Venezuelan President Nicolas Maduro who have not been indicted.

## FIFA (71wiki)

The indictments on 27 May 2015 of nine current and former Fédération Internationale de Football Association (FIFA) officials on charges of racketeering and money-laundering changed the sporting landscape overnight. Suddenly a system of "rampant, systemic and deep-rooted corruption" was brought starkly into global focus.

One outcome of this corruption was that Qatar was awarded the hosting of the 2022 Football World Cup. **Box 3.8** outlines the inclusion of the 2022 World Cup into Qatar's 2030 Vision.

### Box 3.8: Qatar's Vision and the 2022 World Cup (71wiki)

*The aim of QNV **2030** is to "transform **Qatar** into an advanced society capable of achieving sustainable development" by **2030**. The plan's development goals are divided into four central pillars: economic, social, human and environmental development. Strategies undertaken in line with achieving Qatar's Vision included a number of sporting events.*

*The Doha Asian Games of 2004 was the first of such events and was regarded a success. The 2019 World Athletics Championships was another step in this direction.*

*Qatar used behind the scenes activities to influence the FIFA decision to hold the 2022 World Cup in Qatar, the first in the Arab World. Qatar promoted their hosting of the tournament as representing the Arab World, and positioned their bid as an opportunity to bridge the gap between the Arab World and the West.*

*Preparation for the World Cup has seen unprecedented construction activity. A complete new light rail system is being built. At one stage, Qatar had the most Tunnel Boring Machines (TBMs) in operation ever, in one place. Complete new stadiums are being built where the stadium design includes dis-assembly of the upper tiers after the World Cup for donation to countries with less developed sports infrastructure.*

## Russia: Trioka Laundromat (72website)

It is believed that half of Russia's wealth is stashed in offshore tax havens. Leaked data from Troika Dialog – once Russia's largest private investment bank – shows that the bank created at least 75 shell companies in tax havens around the world. When opening accounts in European banks – such as Ukio Bankas in Lithuania (now-defunct), Raiffeisen in Austria and Commerzbank in Germany – the real owners hid behind the paperwork of unwitting Armenian seasonal workers.

These companies channeled at least US$26 billion between 2006 and 2013. Some of this money flowed out of the Troika Laundromat and into the global financial system as clean cash. As a result, Russian oligarchs and politicians secretly acquired shares in state-owned companies, bought real estate both in Russia and abroad, purchased luxury yachts and hired music superstars for private parties. A significant amount of money has been used to purchase real estate in Central London.

## Maldives: Tourism Investments (73video)

In 2016, Al Jazeera revealed that approximately US$1.5 billion was laundered through fake tourism investments in a scheme of astounding simplicity. The money was allegedly transported to the Maldives in cash, approved by the financial authority and transferred to private companies, where it appeared as clean profits from tourism investments.

President Abdulla Yameen (51website)(52news) was voted out of office in November 2018 and was subsequently sentenced to 5 years for money laundering.

## State capture

### South Africa: President Zuma and the Gupta Brothers (74news)(97video)

In what's been described as a "modern coup", the Gupta family took control of South Africa. Through bribing politicians, giving lucrative jobs to President Zuma's children and other ways of buying influence, Ajay, Atul, and Rajesh Gupta captured the state. The Gupta brothers took as much as US$7 billion in government funds. The Guptas also hired and fired government ministers, while the president fired tax officials and intelligence chiefs to protect the family from investigation.

President Zuma has since lost government office and faces corruption and money laundering charges. His successor, President Ramaphosa, vowed to clean up the country, but many officials from the previous administration still remain in positions of authority. The Zondo Commission of Enquiry into State Capture continues and Zuma is still free awaiting trial.

Companies implicated include:

- SAP - the world's largest business software company
- ABB - a major international electrical engineering company
- EOH - the largest technology services organization in Africa
- Denel - a State owned arms manufacturer
- KPMG - a firm offering audit, tax and advisory services
- McKinsey - a worldwide management consulting firm that conducts qualitative and quantitative analyses in order to evaluate management decisions across the public and private sectors.

## Whistleblowing

### Panama Papers (75website)

Following a huge leak from the Panamanian law firm, Mossack Fonseca, the Panama Papers exposed the darkest secrets of the financial industry. The Panama Papers showed that Mossack Fonseca created 214,000 shell companies for individuals who wanted to keep their identities hidden. Behind the shell companies hid at least 140 politicians and public officials, including 12 government leaders and 33 individuals or companies who were blacklisted or on sanction lists by the United States government for offenses like trafficking and terrorism.

Since the exposure, several heads of government have resigned or faced prosecution, at least 82 countries launched formal investigations and Mossack Fonseca closed. Several countries committed to ending financial secrecy, with at least 16 countries and international bodies achieving at least one substantial reform and approximately 23 countries recovering at least US$1.2 billion in taxes.

### Bermuda: Paradise Papers (76website)

In 2017, a secret financial universe was exposed in a huge leak of documents from the Bermuda-based elite legal firm, Appleby. Dubbed the Paradise Papers, the investigation shed light on the widespread use of secretive tax havens by 120 politicians, royals, oligarchs and fraudsters.

The Paradise Papers showed how corporations used these havens to reduce their taxes drastically, and in some cases, commit crimes. Commodities giant, Glencore, is just one company that has been exposed. (see also **paragraph *"Minerals Exploitation"***)

### Spain: the Rajoy government and the Gürtel case (77news)

Over 10 years, the Gürtel case grew into to the biggest corruption scandal in Spain's democratic history, reaching all the way to the president's office. At the centre, the complex scheme funneled illicit donations and bribes to the then-ruling party in exchange for rigged government contracts.

Businessman Correa eventually received a 51-year jail sentence, while a close ally and former treasurer of former president Mariano Rajoy was fined nearly US$50 million. The scheme was discovered by Ana Garrido Ramos, a whistleblower who was also a key witness in this case, contributing to the collapse of the Rajoy government in June 2018.

## Political donations and bribes

### South Africa: Hitachi donation to the African National Congress

The US Securities and Exchange Commission case against Hitachi is outlined in *Box 3.9*.

> Box 3.9: Political donation (78website) (edited)
>
> SEC Charges Hitachi With FCPA Violations
>
> Washington D.C., Sept. 28, 2015
> *The Securities and Exchange Commission (SEC) today charged Tokyo-based conglomerate Hitachi, Ltd. with violating the Foreign Corrupt Practices Act (FCPA) when it inaccurately recorded improper payments to South Africa's ruling political party in connection with contracts to build two multi-billion dollar power plants.*
>
> *Hitachi has agreed to pay $19 million to settle the SEC charges.*

### Brazil: Lava Jato (79news) (98movie)

What began in 2014 as the Lava Jato investigation, or "Operation Car Wash", involving a network of more than 20 corporations – including Brazilian oil and construction giants, Petrobras and Odebrecht – has since grown into **one of the biggest corruption scandals in history.** Involving nearly US$1 billion in bribes and more than US$6.5 billion in fines, the case extends across at least 12 countries in Latin America and Africa. More than 150 politicians and business people were convicted in its wake, including one president, and indirectly, two successors. Odebrecht's fate is described in *Box 3.10*.

Box 3.10: Odebrecht: One of the biggest corporate corruption cases in history (80news)

**What has happened to Odebrecht since?**

*The value of Odebrecht bonds was sliced. The downturn in construction that went alongside this contributed to a downturn in the Brazilian economy and the credit ratings agency Standard and Poor's cut the company to the lowest grading. It has recovered somewhat but is still far from excellent.*

*The investigations have even contributed to economic downturn in Brazil, and Peru shaved almost a point off its growth forecasts for 2017, which the government blamed on "the Odebrecht effect".*

## Minerals exploitation

### Myanmar: Jade (81news)

Myanmar is a tragic example of how rich natural resources are often exploited by the corrupt while causing social and environmental disasters that affect local people. In 2015, a report revealed that corrupt military officials, drug lords and their cronies, had been illegally exploiting jade mines in northern Myanmar and smuggling the stones to China. In total, more than US $31 billion in jade stones were extracted in 2014 alone – the equivalent of half of Myanmar's GDP that same year.

In March 2018, Myanmar submitted Extractive Industries Transparency Initiative (EITI) reports for fiscal years 2014-15 and 2015-16. This shows a move towards transparency.

### Democratic Republic of Congo (82website)(83news)

Commodities giant, Glencore, allegedly bribed the former president of the Democratic Republic of Congo, Joseph Kabila, while it negotiated for mining licenses. (see also **paragraph "Bermuda: Paradise Papers"**)

## Dictators and corrupt politicians

### Nigeria: Abacha (84news)

Sani Abacha was a Nigerian army officer and dictator who served as the president of Nigeria from 1993 until his death in 1998. His five-year rule was shrouded in corruption allegations, though the extent and severity of that corruption was highlighted only after his death when it emerged that he took between US$3 and $5 billion of public funds.

Nigeria is slowly trying to recover the money. US Justice department and UK dependency Jersey have been identified for recovery of some of the money.

It is interesting to note that Abacha ordered the hanging of activist 'Ken' Saro-Wiwa (see **Box 2.10**)

## Peru : Fujimori (85news)

Fujimori presented a clean image to the public during his presidency while using death squads to kill guerrillas and allegedly embezzling US$600 million in public funds. After fleeing to Japan in 2000, he became the first elected head of state to be extradited to his home country, and then tried and convicted for human rights abuses. He was sentenced to more than 30 years in prison.

## Tunisia: Ben Ali (86news)

From 1987 to 2011, President Ben Ali created laws to prevent companies from investing and trading in certain sectors. This allowed him to shut competition out whilst letting 220 family businesses monopolize numerous industries, including telecommunications, transport and real estate. In 2010, these businesses produced 3% of Tunisia's economic output, but took 21% of the private sector profits. The Ben Ali family amassed a fortune of US$13 billion.

The government was overthrown in the Arab Spring of 2011, which was triggered by an unemployed Tunisian setting himself alight.

## Ukraine: Viktor Yanukovych (87news)

Yanukovych and his family fled to Russia in February 2014 after civil unrest. Three years after these tragic events, a Ukrainian court found

Yanukovych guilty of high treason and sentenced him to 13 years in prison in absentia. Former President Viktor Yanukovych and his associates allegedly made US$40 billion in state assets disappear. So far, the Ukrainian government has recovered just US$1.5 billion.

## Equatorial Guinea: President Obiang and family (88news)(53news)

This oil-rich country has the highest per capita income in Africa, but about three-quarters of its population lives in poverty. Since 1979, the ruling Obiang family, along with their cronies, have stolen billions of dollars from the people.

As the most conspicuous and international spender in this kleptocracy, justice caught up with vice president Teodorín Obiang several times. In 2014, the US Department of Justice prosecuted him for money laundering and seized US$30 million worth of assets. In 2017, French authorities found him guilty of embezzlement and confiscated his assets worth US$35 million, while Switzerland seized 24 of his supercars. Obiang is still in power.

## The Gambia: former President Jammeh (89news)

A report by the Organized Crime and Corruption Reporting Project (OCCRP), a non-profit investigative reporting outlet, said that Jammeh and his associates "looted or misappropriated" at least $975m.

## Conclusion

Learning from mistakes is crucial for improvement. Sadly, many companies have not learned from their own mistakes or the mistakes of others, often with irreversible consequences for both the environment and humans.

One would think that industries would learn from the past and reduce the risk of pollution, but the reckless pursuit of profit has pushed companies to be even more destructive. It's important to recall past disasters and corrupt activities to prevent them from recurring. **If such knowledge is not passed on to the next generation, these mistakes will be repeated, possibly with even greater detriment.**

# CHAPTER 4: MEASURING SUSTAINABLE SUCCESS

## Introduction

Although it is difficult to measure real long term success, there are a number of institutional indicators which are used to do so. My particular focus in this chapter will be on those indicators which relate to global warming and the state of Earth. I group them under the main headings of good governance, sustainability, ethical leadership and expecting-the-unexpected.

## Measurement methods for real long term success

*"If you can't measure it, you can't improve it"*

How do we know what the most important indicators are and whether improvements are being made? Indications of improvement should be supported by:

1. baseline status,
2. plans for improvement,
3. actions taken,
4. target dates for actions and
5. whether action targets and dates have been reached.

Measurement includes the use of metrics, reports, lists, certification and awards. I give some examples.

### Metrics

Metrics include indices, percentages, ratios, status, numbers and ratings.

Indices will have changes over time but may not give supporting reasons for improvement or deterioration *(e.g. the Corruption Perception Index for a country)*.

Changes in percentages could imply some improvement *(e.g. a change in the percentage of companies with separate chairman and CEO year-on-year)*.

Ratios could reveal comparisons *(e.g. the amount of sustainable investment to other investment)*.

Status could demonstrate either a yes or no *(e.g. countries that have a democratic political system)*.

Numbers could indicate the rise or fall of a few incidents *(e.g. the number of companies brought before a competitions authority)*.

Ratings could be financial *(e.g. the Moody rating of a country or company)*.

**Other indicators**

Reports give status of the previous reporting period relative to the current reporting period *(e.g.Company Climate Change Financial Disclosure Report)*.

A list of commitments is a tool used to advise what is to be done, by whom and when *(e.g. UN Paris Accord Tracker)*.

Certification indicates that a company has a series of improvement projects to achieve *(e.g. Certification to ISO 14001 Environmental management systems)*.

Awards indicate that a company is committed to improve based on certain predefined criteria *(e.g. ASEAN Corporate Governance Award)*.

## Improvement: cultivating a learning organization

> *"The learning organization is capable of continual regeneration from the variety of knowledge, experience and skills of individuals within a culture that encourages mutual questioning and challenge around a shared purpose or vision."* (1book)

Advocates of the learning organization point out that the collective knowledge of all the individuals in an organization usually exceeds what the organization itself "knows" and is capable of doing, whereas the formal structures of organizations typically stifle knowledge and creativity. They argue that the aim of management should be to inspire processes that unlock the knowledge of individuals, and encourage the sharing of information and knowledge. Every individual then becomes sensitive to changes occurring around them and contributes to the identification of opportunities, risks, and required changes.

The Board of Directors is the brain of the learning organization and the "Fulcrum" of business performance. It draws data from the external environment and business operations and is responsible for making appropriate decisions to take the company forward. Board diversity is essential for cultivating a broader approach to prevent "group think".

The central tenets of organizational learning are that:

• Managers **facilitate** rather than direct

• Information flows and relationships between people are lateral as well as vertical (up and down)

• Organizations are pluralistic, where conflicting ideas and views are welcomed, surfaced, and become the basis of debate

• Experimentation is the norm, so ideas are tried out in action and in turn become part of the learning process.

## Existing institutional performance measurements

Various institutions have established ways in which to measure the 'performance' or 'goodness' of both countries and companies. I've provided some examples.

## Political stability, failed states and misery

Political stability is generally considered ideal in order for businesses to flourish. Yet opportunities exist even where the political environment is less than perfect. Fragile State (2website), Poorest Country (3website) and Misery Index (4website) indicators may give investors pointers for investments, albeit potentially at a higher risk.

Large resource companies weigh the risk of operating in risky environments against their total operation and sometimes choose to take higher risks for certain ventures. Fragile state and poorest country investments are obviously high risk but could offer high returns. The Democratic Republic of Congo (DRC), for instance, is considered one of the five poorest countries and is also regarded as a fragile state, yet Glencore has invested substantially in it's mining sector. Petronas Malaysia has invested in South Sudan, also one of the 5 poorest countries.

The Misery Index is another indicator with Venezuela sadly being regarded as the most miserable country, yet with one of the largest oil reserves in the world. Thailand is regarded as the least miserable and boasts a booming economy. Bhutan is the only country to date to have a Gross National Happiness Index.

On a positive note, DRC, Uganda and Mozambique, regarded as 3 of the 5 poorest states in the world, are currently opening up development of resources that could lift them out of the poverty trap. DRC currently produces 63% of the world's cobalt, while Uganda has oil reserves which are the fourth-largest in sub-Saharan Africa, behind Nigeria, Angola, and South Sudan. Mozambique has extensive offshore gas fields which are currently being exploited.

## Corruption

Corruption is a major stumbling block to development. It tends to be prevalent in undemocratic countries where big international companies bribe corrupt dictators and politicians for concessions to mine and extract oil & gas. The State Operated Enterprises (SOCs) of these countries don't follow accepted contracting practices for award of contracts which allows for backhanders and bribes. Additionally, the top military decision makers are bribed by international arms dealers.

Corruption can be assessed against the Corruptions Perception Index (5website), which was developed by Transparency International (TI). TI scores countries on how corrupt their public sectors appear to be, and also assists companies with investment strategies, for example by suggesting which countries to avoid investing in. Based on expert opinion from around the world, The Corruption Perceptions Index (CPI) uses a scale of zero (highly corrupt) to 100 (very clean). Of the 180 countries assessed in the 2017 index, more than two-thirds scored below 50. In 2018 Denmark and New Zealand were rated as the least corrupt countries while Somalia was the most corrupt.

Transparency International also publishes a Bribe Payers Index (6website). The latest survey of 2011 ranks 28 of the world's largest economies according to the perceived likelihood of companies from these countries to pay bribes abroad. It is based on the views of business executives as captured by Transparency International's 2011 Bribe Payers Survey. Countries perceived to be least likely are Switzerland and The Netherlands, with Russia and China at the other end of the scale. Russia focuses on the oil and gas sector, whereas China concentrates on infrastructure and mining activities, particularly in Africa. Public works contracts and construction projects appear to be the most vulnerable to bribery practices, resulting in inflated prices to pay the bribes.

Other tools for identifying bribery and fraud include Kroll's Anti Bribery and Corruption Bench Marking Report (7website) and the Global Fraud and Risk Report (8website),

## Regulation

Countries need to have basic laws to protect companies and individuals from unscrupulous operators who bribe officials to get competitive advantage . These include laws for the formation of companies, exploitation of natural resources and regulation of stock exchanges.

In many countries, anything under the ground or in coastal waters is state owned. Those wishing to exploit these resources are required to obtain permission from the State and are thus required to pay royalties and/or taxes to the State based on production. **In poorly regulated countries the royalties are either not paid or go into the pockets of corrupt politicians and officials.** To counter this problem, the Extractive Industries Transparency Initiative (EITI) (62website) was formed.

The governing bodies of stock exchanges determine the criteria for listing on national and regional stock exchanges. Non compliance results in fines being imposed or even the delisting of a company. The US Securities and Exchange Commission (SEC) can impose penalties on any company which is quoted on any US Stock Exchange, and, as a result, many multinationals get caught by the SEC for unacceptable practices outside the US.

> The US Securities and Exchange Commission (SEC) fined Hitachi in 2015 and Siemens in 2008 for corrupt practices. (see **Chapter 3**).
>
> The Nigerian Securities and Exchange Commission recently ordered the entire board of an oil and gas company to resign (9news).

## Money laundering

Money laundering by corrupt leaders, drug lords and terrorist groups is a major problem. Many countries have limits on the transfer of funds in order to deter the laundering of money derived from illegal means such as drug smuggling. Offshore tax havens are hotbeds for money laundering, including some Caribbean Island States.

The Basel Anti-Money Laundering Index (10website) is an independent annual ranking that assesses money laundering and terrorist financing (ML/TF) risks for various countries. Published by the Basel Institute on

Governance, it is a composite index made up of highly reliable sources, such as the FATF Mutual Evaluation Reports, as well as a range of individually weighted indicators of corruption, transparency and laws.

In 1989, an inter-governmental body, the Financial Action Task Force (FATF) (11website), was established.

> *"The Financial Action Task Force (FATF) is the global money laundering and terrorist financing watchdog. The inter-governmental body sets international standards that aim to prevent these illegal activities and the harm they cause to society."*

## Investment ratings

Several international rating agencies conduct investment ratings of countries and large organizations, using scales to assess these ratings. These ratings influence the ability of governments and businesses to borrow funds for their operational needs and for specific major investments.

Moody (US) assess long term investments within a credit rating scale, shown in **Table 4.1**. The range spans from 'Minimum credit risk' at the top to 'In default, little prospect of recovery' at the bottom of the scale.

Table 4.1: Moody's Credit Rating Scale (12website)

| Grade | | Long-Term Rating |
|---|---|---|
| INVESTMENT GRADE | Aaa | Smallest degree of risk |
| | Aa1 | Very low credit risk |
| | Aa2 | |
| | Aa3 | |
| | A1 | Low credit risk |
| | A2 | |
| | A3 | |
| | Baa1 | Moderate credit risk |
| | Baa2 | |
| | Baa3 | |
| NON-INVESTMENT GRADE | Ba1 | Questionable credit quality |
| | Ba2 | |
| | Ba3 | |
| | B1 | High credit risk |
| | B2 | |
| | B3 | |
| | Caa1 | Very high credit risk |
| | Caa2 | |
| | Caa3 | |

Standard & Poors (S&P) (13website) conducts, among other things, Credit Ratings and Environmental, Social and Governance (ESG) ratings (14website) (see **Sustainability paragraph**).

Fitch (15website) undertakes ratings similar to Moody and S&P.

Based in Hong Kong, Dagong (16website) undertakes ratings similar to the above US Rating Agencies.

## Corporate governance

Good corporate governance and business ethics are key to the long term investment potential of companies. Various countries have developed Corporate Governance Codes (see **Chapter 1**).

Ethisphere publishes a list of the World's Most Ethical Companies

(17website) where companies are recognized for exemplifying and advancing corporate citizenship, transparency and the standards of integrity. In 2019, 128 companies were listed spanning 21 countries and 50 industries.

It is interesting to note that 3M is included on the list. After the Teflon incident with Du Pont (see **Chapter 3 Learning from History)**, 3M ceased production of Teflon. Kimberly Clark, the makers of Kleenex tissues, is also listed for their efforts in sustainability (see **Box 4.5**).

## Sustainability

Sustainability is essential for the long term survival of any company. RobecoSAM (18website) is an international investment company with a specific focus on sustainability investments and considers economic, environmental and social criteria in its investment strategies.

The Dow Jones Sustainability Index tracks the stock performance of the world's leading companies in terms of economic, environmental and social criteria. It was launched in 1999 as the first global sustainability benchmark and is offered cooperatively by the RobecoSAM and S&P Dow Jones indices. The indices are used by investors who wish to integrate sustainability considerations into their portfolios. Only the top ranked companies in terms of corporate sustainability within each industry are selected for inclusion in the DJSI family and no industries are excluded from this process. Selected criteria are:

1. Corporate governance
2. Supply chain management
3. Risk and crisis management
4. Operational eco-efficiency
5. Climate strategy
6. Releases to the environment
7. Social Impacts on communities
8. Occupational health and safety
9. Labor practice indicators and human rights.

Thai Oil PLC was Energy Industry Group Leader in 2019 (19website).

Standard & Poor (S&P) Environment Sustainability & Governance (ESG) Index (20website) provides investors with a view of companies' ESG profiles. Profile factors include:

- Environment: green house gas (GHG) emissions, waste & pollution, water & land use
- Social: workforce & diversity, safety management, customer engagement and communities
- Governance: structure & oversight, code & values, transparency & reporting, cyber risk & systems.

The index is based on S&P Dow Jones Indices ESG Scores.

S&P Fossil Fuel Free Index measures the performance of companies that do not own fossil fuel reserves, while over weighting and under weighting companies based on their levels of carbon emissions.

## Awards

Countries and regions present awards for excellence in management of companies. These include:

The Malcolm Baldrige National Quality Award (21wiki) which recognizes US organizations in the business, health care, education, and nonprofit sectors for performance excellence, based on the Baldrige Performance Excellence Program.

The EFQM Excellence Award is run annually by the European Foundation for Quality Management (EFQM) and is designed to recognize organizations that have achieved an outstanding level of sustainable excellence, based on assessment against the EFQM Excellence Model. The EFQM Excellence Model (22website) is the most widely used continuous improvement tool in the world and can be applied by any organization regardless of size or sector.

The Australian Business Excellence Awards (23website) are recognized as Australia's premier business awards. They present an opportunity for a broad range of businesses and organizations from across Australia to celebrate and showcase their excellence against internationally recognized business principles.

The Association of South East Asian Countries (ASEAN) (24website) comprises Vietnam, Cambodia, Laos, Thailand, Philippines, Malaysia, Myanmar, Brunei and Singapore. Improvements in the regulation of the region's capital market were triggered by the Asian financial crisis (Asian Flu) in 1997 and outcomes include the formation of the ASEAN Corporate Governance Scorecard (see **Box 1.7**) and subsequently the ASEAN Corporate Governance Awards.

## Financial assessments

Financial assessments include the following:

Audited Company Annual Reports are required to be produced by all public companies for their shareholders. These reports must include all aspects of performance of the business including its financial well being. Profit after tax (PAT) and return on investment (ROI) are key elements.

Gross Domestic Product (GDP) is a monetary measure of the market value of all the final goods and services produced annually within a country. The US Central Intelligence Agency lists the GDPs for all countries (25website). It should be kept in mind that the GDP of a country is just one small element of a much bigger context and its importance can be over stressed when assessing a country's wealth. For instance, Brunei and Qatar are of the richest countries according to GDP. Yet this is based on oil economies and very small populations, whereas the GDP of a country with no natural resources, such as Singapore, is earned by its people. Venezuela, on the other hand, boasts some of the highest oil reserves in the world, yet an inept government has destroyed their economy.

## Social progress (26website)

Traditional measures of national income, such as GDP per capita, fail to capture the overall progress of societies. A new index being adopted is the Social Progress Index (SPI). This index captures outcomes related to all 17 Sustainable Development Goals and is a comprehensive snapshot of a country's overall progress towards the achievement of the goals. In Paraguay, the central government has officially adopted the index as part of the National Development Plan, doubling its budget for nutrition programs as a result of the priorities highlighted by the data. In Brazil, multinational corporations like Coca-Cola, Natura and Fiat-Chrysler are using customized indexes to ensure their supply chains are socially and environmentally sustainable.

The key elements of the index are:

A. Basic Human Needs

> Nutrition and basic medical care
> Water and sanitation
> Shelter
> Personal safety

B. Foundations of wellbeing

> Access to basic knowledge
> Access to information and communications
> Health and wellness
> Environmental quality

C. Opportunity

> Personal rights
> Personal freedom and choice
> Inclusiveness
> Access to advanced education.

## Other assessments

Fortune magazine publishes lists of the 'world's most admired companies' (27website). For their '50 most admired companies overall' list, Fortune's survey asks business people to vote for the companies that they admire the most, from any industry. The listing is based on innovation, people management, use of corporate assets, social responsibility, quality of management, financial soundness, long-term Investment value, quality of products/services and global competitiveness.

IMD produces IMD World Competitiveness Ranking (WCR) and World Talent Ranking (WTR) (28website). IMD is an independent business school, with Swiss roots and a global reach. In 2019, Singapore was ranked first in WCR while Switzerland was ranked first in WTR.

International Finance Corporation (IFC) (29website) has produced Performance Standards on Environmental and Social Sustainability which have become globally recognized good practice in dealing with environmental and social risk management. More than 90 banks and financial institutions have voluntarily adopted the Equator Principles, which are based on IFC's Performance Standards.

Organization for Economic Co-operation and Development (OECD) (30website) has a wealth of statistics by country and industry. The OECD is an intergovernmental economic organization with 36 member countries, founded in 1961 to stimulate economic progress and world trade.

## Grouping measurements for sustainable success

As discussed in **Chapter 1**, key factors for success are grouped as follows:

1. Good governance
2. Sustainable business model
3. Ethical leadership and eradication of corruption
4. Expecting the unexpected.

I list some examples of measurements, categorized into these four key areas. Measurement types are shown in *italics*.

## 1 Good governance

Good governance is introduced in **Chapter 1**.

### Fortune's 50 most admired companies *(list)*

Fortune magazine publishes what they call 'lists of the world's most admired companies'.

### Investment rating agencies *(rating)*

Moody's etc. give financial ratings for investment purposes.

### Corporate governance rating *(index)*

Governance ratings are popular in Asia and Africa.

For companies, The ASEAN Corporate Governance Scorecard is described in **Box 1.7**.

For countries in Africa, the Mo Ibrahim Governance rating (31website) is used.

### Business excellence awards *(status)*

Various awards for Business Excellence are given in different countries. However, only ASEAN awards companies for Corporate Governance.

### Extent of democracy *(status)*

Progress towards democracy may give an indication of relative stability and rule of law. Freedom House rate countries free, partly free and not free (32website).

Checks and balances are required to prevent autocratic behaviour by a Head of State or Chief Executive of a Company. Part of good governance is the separation of the roles of government institutions: Executive/Administration, Legislature and Judiciary, as is the case in a strong democracy. In the case of companies, the separation of Chairman of the Board and Chief Executive Officer (CEO) of the Company is essential. (see **Box 1.19**).

Democracy, however, does not directly link to good governance of countries and companies.

## Corruption Perception *Index* (CPI)

CPI is high in countries where corruption is low. The stability necessary for establishing and operating a business is found where the CPI is high.

## 2 Sustainable business model

At the beginning of this chapter, I introduced a number of indicators for a sustainable business model. The Dow Jones Sustainability Index (DJSI) and S&P ESG Index are some of the measurements available. In addition to these the S&P Fossil Fuel Free and ESG Factor Indices are useful, while the Equator principles framework facilitates the assessment of environmental and social risk.

## Global warming countdown *(reports)*

Inline with the Paris Accord countries are committed to keep global warming to a maximum increase of 1.5°C by 2030.

According to the BP report of 2019 (49website), there are indications that global warming is increasing at a higher rate than expected since many businesses are operating in a 'business as usual' mode. Despite this grave forewarning, the Covid-19 pandemic reduced global energy demand by 8% in 2020 (62website).

Support from big energy resource companies is essential to move towards a circular economy. The largest contributors to global warming through the supply of the energy providing raw materials, such as oil, gas and coal, need to be restrained. An analysis identifying those who have made the most substantial contributions since 1965 is shown in **Box 4.1**.

## Box 4.1: Global Warming Contributors (33news)

*The top 20 companies have contributed to 480bn tonnes of carbon dioxide equivalent since 1965*

Billion tonnes of carbon dioxide equivalent

| Company | Value | Company | Value |
|---|---|---|---|
| Saudi Aramco | 59 | PetroChina | 16 |
| Chevron | 43 | Peabody Energy | 15 |
| Gazprom | 43 | ConocoPhillips | 15 |
| ExxonMobil | 42 | ADNOC | 14 |
| National Iranian Oil Co | 36 | Kuwait Petroleum Co | 13 |
| BP | 34 | Iraq National Oil Co | 13 |
| Royal Dutch Shell | 32 | Total SA | 12 |
| Coal India | 23 | Sonatrach | 12 |
| Pemex | 23 | BHP Billiton | 10 |
| Petróleos de Venezuela | 16 | Petrobras | 9 |

These top 20 companies have contributed 35% of all carbon dioxide and methane since 1965. So far only BP (34news) and Royal Dutch Shell (35website) have committed to reduce their effect on global warming and are taking action.

### Task Force on Climate-related Financial Disclosure (TCFD) Status Reports (36website)

In June 2017, The Task Force on Climate-related Financial Disclosures (TCFD) released its final recommendations, which provide a framework for companies and other organizations to develop more effective climate-related financial disclosures through their existing reporting processes.

Investors need accurate and effective information on how companies have prepared or are preparing for a lower-carbon economy. Those companies that meet this need may have a competitive advantage.

Valid climate-related financial disclosures are necessary based on TCFD recommendations.

Network for Greening the Financial System (NGFS), issued recommendations aimed at facilitating the role of the financial sector in achieving the objectives of the Paris Agreement. One of these recommendations encourages all companies issuing public debt or equity as well as financial sector institutions to disclose in line with the TCFD recommendations.

> *"Our review of over 1,000 companies showed that, for some recommended disclosures, the percentage of companies disclosing information increased up to nearly 15% over a two-year period."*
> 2019 TCFD Status Report

## Customer preferences *(reports)*

Customer Preferences (reports) are key to influencing the move towards sustainability. For instance, in the UK and some other European countries, a consumer can choose power and heating suppliers who adopt more sustainable approaches to business.

Purchasers are also becoming more selective about how their food is produced. Coffee lovers can now choose coffee from responsible growers. (Starbucks is leading in this field.) Organizations such as Fairtrade (37website) are improving the lot of small scale farmers. **Box 4.2** outlines what Fairtrade does.

### Box 4.2: What Fairtrade does (38website)

> *"Fairtrade is a simple way to make a difference to the lives of the people who grow the things we love. We do this by making trade fair.*
>
> *Fairtrade is unique. We work with businesses, consumers and campaigners. Farmers and workers have an equal say in everything we do. Empowerment is at the core of who we are.*
>
> *We have a vision: a world in which all producers can enjoy secure and sustainable livelihoods, fulfill their potential and decide on their future.*

*Our mission is to connect disadvantaged farmers and workers with consumers, promote fairer trading conditions and empower farmers and workers to combat poverty, strengthen their position and take more control over their lives."*

Sustainability in the food and clothing supply chains has become a key factor for success. Large supermarkets have a big influence over their suppliers. Ethical business practices sustain a good long term relationships and some companies advise their suppliers on ways to retain high quality standards which has an added advantage as a selling tool to both investors and customers. Woolworths Holdings South Africa's Good Business Journey (39website), outlined in **Box 4.3**, is a good example.

### Box 4.3: Woolworth's Good Business Journey

*'Our Vision is to be one of the World's most responsible retailers.'*

*Categories are:*

- *People and transformation*
- *Social Development*
- *Health and welfare*
- *Sustainable farming*
- *Ethical sourcing*
- *Waste*
- *Water*
- *Energy and climate change.*

## Shareholders: institutional investors and other investment groups (ratios)

Institutional investors tend to be pension funds and insurance companies requiring long term investment returns. Some of these investors now state ESG requirements in their investment preferences.

While certain investors are tending to move away from investments in fossil fuels, others are placing pressure on companies to reduce their carbon footprint. Some large fossil fuel companies are investing in carbon capture, use and storage (CCUS) technology and related fields. (See **Chapter 5 paragraph "Action being taken"**)

*Examples*

> 1. Certain investors refuse to invest in big tobacco.
>
> 2. Islamic investors, through Islamic Banks, have very specific areas of investment
>
> 3. The Norwegian Sovereign Fund has divested from logging companies.

*Irony*

> *Rupert International, a multinational tobacco company, invests in private health care in South Africa. Could this be the result of smokers suffering from lung cancer?*

**Some investors invest exclusively in pure sustainable companies such as solar and wind energy.** Blackrock (41website), the world's largest asset management firm, has announced bold steps to integrate sustainability into its investment portfolios with a focus on climate change.

## Company sustainability reports and goals *(reports)*

As noted in an earlier paragraph, the Task Force on Climate related Financial Disclosure (TCFD) gives a framework for these financial disclosures.

Ikea has set a good example for what companies can achieve. An extract of it's 2018 Sustainability report is shown in **Box 4.4**.

> Box 4.4: Ikea 2018 Sustainability Report extract (42website)
>
> ➢ *New circular design principles and services – designing new products to*

be re-purposed, repaired, reused, resold or recycled

- More renewable and recycled materials – during FY18, 60% of the IKEA range was based on renewable materials and 10% contained recycled materials. The ambition is 100% renewable and recycled materials by 2030
- Increasing renewable energy - 18,240 solar panels installed on IKEA Industry production unit in Portugal - enough to power 2,700 homes
- The number of home solar customers grew and customers saved money on their energy bills. Today, home solar is offered in six IKEA markets
- Sourcing responsibly - 100% of cotton (same as FY17) and 85% (up from 77% in FY17) of wood from more sustainable sources
- Developing new innovative products such as the MISTELN – a water saving mist nozzle that can reduce water use by more than 90%
- Launch of three collections in partnership with social enterprises
- More plant-based food – veggie hot dog was launched globally, it has 85% smaller climate footprint than the normal hot dog
- Commitment to phase out single-use plastic products from the IKEA range by 2020.

Kimberly-Clark (43website), the manufacturers of Kleenex tissues, claim that their tissues are made of virgin fibre which is sourced from certified sustainable forests. ***Box 4.5*** outlines Kimberly-Clark's sustainability goals.

## Box 4.5: Kimberly-Clark progressing toward sustainability goals
## July 4, 2018 (extract) (44news)

*Kimberly-Clark's 2017 Global Sustainability Report outlines the company's strategies and results in greater detail, and is organized and presented in accordance with the Global Reporting Initiative (GRI) Sustainability Reporting Standards, Core level. Key outcomes include:*
*• Social Impact – In the first two years, the lives of 4.3 million people in need worldwide have benefited through programs that increase access to sanitation, help children thrive and empower women and girls.*

- *Forests & Fibre* – Sourced 89 per cent of the fibre used in tissue products from environmentally preferred sources, including FSC® certified virgin fibre and recycled fibre.
- *Waste & Recycling* – Further increased the amount of waste diverted from landfills (includes both manufacturing and non-manufacturing facilities) to 95 per cent.
- *Energy & Climate* – Lowered the absolute greenhouse gas (GHG) emissions by 18 per cent (from 2005 baseline) through improved energy efficiency and increased use of alternative energy sources.
- *Supply Chain* – Invested in watershed analyses, as well as water treatment and recycling technologies, to further reduce water consumption in water stressed areas.

Additionally, Kimberly-Clark announced expanded programming in several key areas.

The company began measuring and reporting on Scope 3 GHG emissions – the indirect emissions that occur both upstream and downstream in the value chain – and will establish reduction targets later this year. The company worked with the World Resources Institute and World Wildlife Fund to identify the company's primary sources of Scope 3 emissions.

This work comes on the heels of a major commitment in 2017 to annually purchase approximately 1,000,000 megawatt hours of renewable wind energy, equivalent to about one-third of the electricity needs of Kimberly-Clark's North American manufacturing operations.

## ISO 14001 Environmental Management Systems compliance (certification)

This global standard ensures continuous improvement in environmental management. One of the few hotels certified to ISO 14001 is described in **Box 4.6**.

### Box 4.6: Application of ISO 14001 - *Shangri La Resort Hotels* <u>(45website)</u>

The Shangri La is the only hotel group in South East Asia listed in the Dow Jones Sustainability Index. 55 hotels in the group are certified to

*ISO 14001.*

*Achievements obtained by applying this standard include:*
- *5.6% reduction in GHG emission intensity since 2015*
- *water consumption intensity at same level as 2015.*

*Various project are undertaken including:*
- *Care for people project*
- *Care for nature project including reef coral, nature and turtles*

*Innovative solutions include:*
- *Glass water bottling*
- *Rainwater harvesting*
- *Use of solar power*
- *Incorporation of composting and herb gardening to manage food waste.*

## UN Paris Accord tracker *(lists)*

Extracts from 2020 Climate Progress Tracker Tool are outlined in **Box 4.7.**

### Box 4.7: 2020 Climate Progress Tracker Tool (extracts)

*Energy (46website)*

- **Chile** *vowed to go carbon neutral by 2050 and is closing eight coal-fired power stations over the next five years as part of a plan to switch entirely to renewable energy by 2040*
- **Great Britain** *hit a new power milestone – lasting for a fortnight without using any coal power to generate electricity for the first time since the industrial revolution*
- *Renewables overtook coal as* **Germany**'s *main source of energy for the first time in 2018*
- *The* **Indian states** *of Gujarat and Chhattisgarh announced they will end construction of coal-fired power plants*
- *In* **Germany**, *sun, wind, water and biomass produced more electricity in the first seven months of 2019 than coal and nuclear power combined*
- **Portugal** *can boast the world's lowest-cost solar photovoltaic contract (US$16.54/MWh) The Portuguese auction awarded 1.15 gigawatts of*

capacity
- The world's biggest back up battery in South **Australia** by Tesla had performed faster, more accurately, more reliably and with more versatility than expected since it was turned on in November 2017 and is expected to earn more than $25 million in revenues (a third of its construction costs) in its first year
- In 2017, renewables accounted for the majority of all new power-generating capacity added in the world
- The cost of lithium-ion batteries fell by 85% between 2010 and 2018
- Glencore, the largest supplier of thermal coal to the international market, said it will cap production for the foreseeable future at around current levels of 145 million metric tons a year.

*Companies (47website)*

*Cement*

- Vattenfall in Germany is planning the world's first zero-carbon cement plant
- Heidelberg cement company in Germany aims to produce CO2-neutral concrete by 2050
- In India, Dalmia Cement, Ambuja Cement, and Shree Cement are now producing the lowest carbon intensity cement in the world – the carbon footprint at its group level is 20% less than the global cement average

*Steel*

- German steelmaker Thyssenkrupp plans to phase out $CO_2$-intensive coke-based steel production and replace it with a hydrogen-based process by 2050

*Wind turbines*

- Danish company Orsted has installed more than 1,000 offshore wind turbines - more than any other company in the world

*Food/ Household*

- Mars is committed to making its roughly 150 factories worldwide carbon neutral by the year 2040
- Danone is committed to becoming carbon neutral across its full value-chain by 2050
- Nestlé, the world's biggest food company, will cut greenhouse gas

emissions across its operations and supply chain to net zero by 2050
- Unilever is to become carbon negative by 2030

Electronics/Media

- Sony is working to reduce its environmental footprint to zero by 2050

Furnishings

- The IKEA Foundation has announced plans to spend another €300 million on climate-related programs between 2020 and 2023.

"I would have a hard time even understanding how you can be successful as a business if you cannot mirror the society that you serve in the first place." Paul Polman, CEO, Unilever (48website)

## Measuring carbon footprints *(number)*

A company's carbon footprint is a useful tool to indicate the rate of improvement. Some companies measure their supply chains carbon footprint, resulting in a far more accurate representation of the effect they're having on the environment. Large supermarkets are moving to more local suppliers to reduce their carbon footprint. Woolworths, a clothing and food retailer in South Africa produces an annual carbon footprint for the company (50website).

The BP Statistical Review of World Energy publishes the per capita carbon footprint by country (49website) with Singapore as the highest at 633 Gigajoules per capita and Bangladesh the lowest at 9 Gigajoules per capita.

## Excellence in transformation business awards (51website) *(status)*

The Financial Times and International Finance Corporation (IFC) have an annual awards event which rewards those members of the private sector which best meet the UN Sustainable Development Goals (SDGs). Main categories are urban infrastructure, health wellness & disease prevention, education knowledge & skills, and food, water & land. Some innovative ideas emanating from this event are listed in **Box 4.8**.

Box 4.8: FT/IFC Awards examples

- *Lithium Electric Vehicles (EVs) in India*
- *Gravity drip Irrigation in Israel*
- *Localised solar-hybrid mini-grids in Nigeria*
- *Digital Agriculture Platform for Small-Holder Farms in Nigeria*
- *Self-adhesive label sensors that monitor a food product's time and temperature history and change colour to give a clear visual indication of the product's remaining shelf life*
- *Data Intelligence for the Agricultural Value Chain in Latin America*
- *Compostible flexible packaging*
- *Various AI based inventions.*

Banking awards (52website) *(status)*

Euromoney have annual excellence awards with categories including the World's Best Bank for Corporate Responsibility and World's Best Bank for Sustainable Finance.

## 3 Ethical leadership and eradication of corruption

The subject of ethics is introduced in **Chapter 1**.

The World's Most Ethical companies are listed by Ethisphere. (see **paragraph *"Existing institutional performance measurements"***).

Boards *(mix and roles)*

A mix of experience and stakeholder representation on a Board of Directors (53website) can also give an indication of the company's commitment to sustainability and ethical leadership. Representation should include gender and cultural diversity. Norway, for example, has set targets for gender equality at board level.

Separation of roles of chairman of the board and chief executive of the company (54news) promotes ethical leadership. A push by corporate governance experts, shareholders and, in some cases, regulators to untangle the chairman and chief executive positions at US public companies is gaining traction. The percentage of S&P 500 companies whose chief executives also serve as chairman was reduced to 45% in 2018, compared to 48% the year before. These roles are traditionally separated elsewhere.

*Examples of conflict of interest with a single Chairman and CEO*

> 1. Tesla Chief Executive Elon Musk relinquished his position as Chairman of the electric car maker to settle a lawsuit by the Securities and Exchange Commission, which alleged he made misleading statements to investors via Twitter *(40news)*.

> 2. Renault separated its CEO and Chairman after Carlos Ghosn, who held both positions, was arrested in Japan on allegations of financial misreporting at partner company Nissan *(63news)*.

## Reputation *(incidents)*

Examples of past corrupt practices are discussed in **Chapter 3**. Examples of loss of reputation are given in **Chapter 1**.

## Money laundering *(incidents)*

Money laundering primarily affects the reputation of banks (see **Chapter 2 paragraph "Banks and money laundering"**). Whistleblowing on money laundering activities destroys reputations (see **Chapter 3 paragraph "Lessons to be learned from major corruption"**).

Improvement in legislation and reciprocal tax & laundering agreements give some illustration of success. Statistics on the prosecution of offenders is indicative of the commitment of governments.

The number of investigative journalist and whistleblowing reports also gives an indication of improvement (or decline) in this field.

## Monopolistic practices *(cases)*

In the software industry, the Microsoft Windows operating system for personal computers is a virtual monopoly while Google has 70% of search engine market. In the aircraft industry, Boeing and Airbus are duopolies in the supply of wide body jet aircraft.

When companies are so far ahead of their competition it is difficult to envisage that they will fail. IBM had a virtual monopoly on mainframe computers, but then unsuccessfully tried to monopolize the personal computer market and eventually sold out to Lenovo. Kodak and Fiji had the majority market for cameras and film which was destroyed by the development of digital cameras.

Monopolies drive up prices. National competition boards or commissions are necessary to reduce price collusion and other negative effects of monopolies and duopolies. Statistics of competition board rulings on monopolistic practices of companies may be of value. Recent cases that have come before the European Union Competition Commission (55website) include:

1.Microsoft inter-operability
2007 – European Court of First Instance confirms Microsoft has abused its dominant position by refusing to supply inter-operability information to rival server vendors and tying Windows Media Player to Windows.

2.Microsoft choice of web browser
2009 – Microsoft is bound by law to enable Windows to run a variety of web browsers in the EU – and to allow computer manufacturers and users to turn off Internet Explorer.

3.Intel
2009 – Intel abused its dominant position on the X86 central processing unit (CPU) market by granting anti-competitive rebates and making payments to delay / stop deployment of competing products. The Commission imposed a €1.06bn fine and obliged Intel to correct the situation.

## Gross Domestic Product (GDP) vs. Social Progress *Index* (SPI)

The dilemma of how to share wealth fairly among peoples of a nation is complicated. GDP does not measure all aspects of a healthy country. The Social Progress Index (SPI) could be the trend in future to ensure that economic growth is no longer the only key indication of the wellbeing of a country.

## National leadership *award*

Former Heads of State in Africa are awarded for excellence in leadership. (see **Box 2.2**).

## Corruption *(index)*

The subject of corruption is introduced in **Chapter 1** and the elimination of corruption is discussed in **Chapter 3** . The enforcement of anti-corruption legislation and the application of codes shows the level of success.

Singapore, under the leadership of Prime Minister Lee, advocated a total intolerance to corruption and initiated a Corrupt Practices Investigation Bureau (CPIB) that has produced encouraging results.

The annual Global Fraud Survey (57website), commissioned by Kroll and carried out by the Economist Intelligence Unit, is a very informative overview of worldwide fraud.

The Corruption Perceptions Index (56website) was created in 1995 by Transparency International. Each year about 200 countries are ranked according to the perception of the level of corruption in the country.

The Corruption Perception index for 2018 shows a decline in confidence in the democratic process. When politicians tell the people what they want to hear just to get into power and then change track, democracy is under threat. Buying votes is common in certain countries, even to the extent of giving free food to the very poor as promised by Madero in Venezuela.

There appears to be a worrying phenomenon among the voting population in a number of countries who are tiring of corrupt politicians and so are tending to move toward right wing policies and attitudes. Greece's new conservative party 'New Dawn' is gathering momentum, while Brazil has a far right President. Donald Trump was elected to the US presidency by the 'right white'.

The downside of moving to the right is that there is an inclination for xenophobia and religious intolerance to increase and so does the abuse of indigenous tribes, as seen in Sarawak Borneo and Brazil (see **Chapter 3**).

The devastating effect of corruption on the economy of a country is illustrated in **Box 4.9**.

### Box 4.9: Corruption at the top

*The Electricity Supply Commission is a State Owned Enterprise with a power generation monopoly in South Africa. The majority power generation is currently from large coal fired power stations, a major contributor to global warming. To expand power generation, the government, in its wisdom, decided to obtain the bulk of its power from nuclear power stations. They decided to tie up with the Russians, who offered the 'best deal' for the senior politicians and their lackeys at the time. The Finance Minister was against the deal which had been negotiated by the Minister of Energy. The Finance Minister was replaced.*

*Fortunately, the President, Jacob Zuma, was replaced in February 2018 and the new President, Cyril Ramaposa, and his appointed Finance Minister, were able to cancel the Russian deal.*

*If the deal had gone through, the country could have been* **bankrupted***.*

*Nevertheless, the rolling blackouts continue. Pressure is on privatizing electricity generation but the labour unions do not want Eskom broken up or any power generation outsourced. In the interim, Eskom is technically bankrupt and continues to bleed the government for funds.*

*In spite of union opposition, Eskom is going ahead with awarding packages for wind and solar power generation to private contractors.*

Actions aimed at combating corruption need to be assessed with respect to progress in implementation and enforcement of anti-corruption laws. These could include:

1. Adopting ISO 37001 Anti bribery management systems

2. Applying Transparency International (TI) counter bribery business principles

3. Complying with Organization for Economic Cooperation and Development (OECD) principles

4. Applying the Extractive Industries Transparency Initiative (EITI) standards for natural resources management.

## 4 Expecting the unexpected

### Scenario planning

Is there any point in scenario planning for the next 30 years? One could counter with the old adage "A plan, even a bad plan, is better than no plan at all". Clem Sunter gave three scenarios for the future of South Africa at the end of apartheid: the high, middle and low roads. (This provoked leaders in South Africa to avoid the low road at all costs). These alternatives could also be used for best, most likely, and worst case scenarios when contemplating the survival of a company. An example is global warming graph in *Figure 6.1*.

In risk management jargon, *known unknowns* are events that have happened but it's not known if or when they will happen again. Past *known unknowns* have to be taken into account when developing different scenarios for the future of the business. The Covid-19 virus pandemic is a great such example. Previous viruses such as the Spanish Flu of 1918, and the recent SARS virus did not have long-term consequences on the global economy, whereas it's expected that the Covid-19 virus may well lead to a long term recession.

"Black swans" are *unknown unknowns* in risk terminology. Again, previous "black swans" have to be noted when developing a range of scenarios for the future of the business. Some examples of unexpected incidents since the 1970s are shown in **Box 4.10**.

## Box 4.10: Major World Events since 1970s

*This is a list of recent major world events that could not have been foreseen:*

*1973: World Oil Crisis as a result of OPEC shutting off their supply of oil to the world*
*1977: Tenerife Airport Disaster: the deadliest aircraft crash in history*
*1990: Saddam Hussain's invasion of Kuwait, resulting in a major oil crisis*
*1998: Asian Financial Crisis, causing financial hardship in these countries for many years*
*2000: Technology Bubble Crash on the New York Stock Exchange, resulting in financial loss to many people*
*2001: September 11 terrorist attack on the New York Word Trade Center and the Pentagon, and the subsequent economic downturn*
*2002: Enron, WorldCom, and other accounting frauds that ended in the SOX legislation in the US*
*2008: Derivative-based financial meltdown, leading to some of the biggest banks in the world going bankrupt*
*2011: Japanese Tsunami, causing multiple nuclear power station meltdowns*
*2010 US Gulf Coast Oil Spill: the biggest ever*
*2014: Three Malaysian aircraft losses in less than 1 year in unrelated incidents*
*2019-20: Covid-19 Virus Pandemic.*

## Nuclear power *(lessons learned or not)*

Nuclear power has long been touted as being "fail safe" but the Chernobyl, Three Mile Island and Fukushima disasters have shown otherwise. Nuclear meltdowns cause long term physical damage to our world and its inhabitants. **Box 4.11** illustrates the unwillingness to listen to advice.

Box 4.11: *Tokyo Electric Power Company (TEPCO) Fukushima Disaster*

In 2011, three nuclear power reactors were permanently damaged due to meltdowns as a result of a tsunami in the worst nuclear disaster since Chernobyl in 1986. These reactors, commissioned between 1971 and 1976, were protected by a 5.7 meter high sea wall.

*2008 tsunami study*

A 2008 in-house study identified an immediate need to better protect the facility from flooding by seawater, stating the possibility of tsunami-waves up to 10.2 meters. Head office officials insisted that such a risk was unrealistic and did not take the prediction seriously. Just three years later, waves in excess of 30 metres were recorded in the so-called Great East Japan Earthquake.

*Others did something (58video)*

Yet others invested in resistance to a potential tsunami. In the 1970's Fudai village (population 3000) invested $30 million (current value) in a 15 meter high floodgate. The 2011 tsunami caused minimal damage to the village while neighbouring villages were destroyed.

## Cyber war *(reports)*

Cyber attacks on companies and countries are increasing and companies need to take active measures to minimize these risks. Saudi-Aramco, the UK National Health Service and Baltimore City business activities were severely disrupted by such attacks (see **Chapter 2 paragraph 'cyber criminals'**) .

## Radicalism *(risk factor)*

Radicalism is threatening business development in certain areas.

Employees of a company, who were building an airstrip for an off-shore LNG plant which could dramatically accelerate Mozambique's economic development, were murdered by radicals in norther Mozambique. Mozambique could remain one of the poorest countries in the world should these radicals continue to terrorize and control the area.

## Viruses and disease *(emergency response planning)*

The Spanish flu of 1918 was a warning to the world of the potentially devastating effects of a viral pandemic. The recent SARS and MERS epidemics were successfully contained, but the Corona (Covid-19) pandemic is the worst the world has known since the Spanish flu. It is interesting to note which countries prepared for the possibility of another epidemic after SARS. Taiwan, Hong Kong and Singapore put their plans into effect to prevent the spread of Covid-19, in a very short time based on their planned strategies. In the USA, President Barak Obama set up a 'strategic planning group' which Donald Trump dismantled the moment he came in power.

Antibiotics are less effective due to the misuse and bacteria becoming more resistant (61news).

## Financial markets

The world financial crash of 2009 has exposed how vulnerable financial markets are. The 2020 world financial crash due to the Covid-19 virus is expected to have a dire long term effect on the world economy.

As discussed in **Chapter 3**, big banker greed and lack of governmental controls led to the collapse of the US financial system and the subsequent fallout on banking around the world. Iceland was badly afflicted by this as well as, but, unlike the US, many of their bankers were punished.

The 2020 financial collapse due to the Covid-19 pandemic has changed the financial world for good and killed '**business as usual**' practices. Companies need to think of a new business model for a sustainable future.

## Natural disasters

A tsunami in December 2004, destroyed parts of Aceh in Indonesia, Phuket in Thailand as well as a large part of Sri Lanka, killing thousands of people. The 2011 Tōhoku earthquake and tsunami in Japan caused 3 nuclear power station meltdowns (see **Box 3.2**).

Turkey experiences many earthquakes and shoddy construction work on buildings and infrastructure projects increases the death rate.

Bali's tourist-based economy was under threat in 2017 when Mount Agung erupted stopping all travel to the country and creating lasting economic damage.

Early in 2019, Mozambique had a major flood in the Beira region. Besides killing thousands, power supply to South Africa from the Cahora Bassa Hydro Power Station on the Zambezi river caused increased strain on the South African power grid causing increased rolling blackouts. This was on top of Eskom's mismanagement and poor maintenance of power generating plant.

Early in 2019 bush fires in Venezuela caused a total blackout of electricity in the country. This was probably the 'last straw that broke the camel's back' with respect to Venezuela's electricity supply system.

The US Gulf Coast gets flooded more and more frequently each year causing refineries in the region to shut down.

## Global warming: the tipping point

We have 10 years to limit global warming to a 1.5 degree increase (59website). The alternative is expected to be catastrophic for the world. A tipping point (60book) could be reached where climate change may suddenly accelerate. Indicators are the Siberian arctic forest fires, the Amazon rain forest fires, melting Alaskan permafrost, English peat bogs drying out and mangroves being eradicated.

## Conclusion

Success in moving towards a sustainable world economy can be measured in a variety of ways. Although not perfect, the categories and measurements outlined in this chapter give an indication of what is being done to move towards a cleaner, greener future. The rate of success has to be accelerated to achieve the goals of the Paris Accord.

# CHAPTER 5: GLOBAL WARMING AND THE STATE OF EARTH

## Introduction

*"140 million people will be displaced by 2050 due to global warming"* [(1website)](#)

The world's leading climate scientists have warned that **there are only a dozen years left for global warming to be kept to a maximum of 1.5C**, beyond which even half a degree will significantly increase the risk of drought, floods, extreme heat and poverty for hundreds of millions of people [(131website)](#).

This chapter gives a brief description of the current effect of climate change on the world, provides evidence of global warming and indicates the degree of deterioration in the state of Earth, and what actions are being taken to slow the process.

*"The world urgently needs to put a laser-like focus on bringing down global emissions. This calls for a grand coalition encompassing governments, investors, companies and everyone else who is committed to tackling climate change."* [(82website)](#)

Swedish Professor Johan Rockström's research has found nine "planetary boundaries" that can guide us in protecting our planet's many overlapping ecosystems [(113tedtalk)](#). While some of the world's leaders are focusing on climate change, we also need to be cognizant of the other planetary boundaries which are:

Biodiversity loss

Land conversion

Nitrogen and phosphorus loading

Ocean acidification

Chemical pollution

Freshwater withdrawals

Ozone layer depletion

Air pollution.

These all have ecological ceilings which are the approximate boundary between sustainable and unsustainable use of ecological processes: the interactions between plants, animals and the non-living components of the environment like climate or rocks. Experts believe that we have already exceeded the ecological ceiling for climate change, biodiversity loss, land conversion and nitrogen and phosphorus loading.

## The effect

There is now overwhelming evidence that the state of Earth has deteriorated considerably due to human activity. I briefly outline evidence of this effect and the main culprits perpetuating the damage.

These are examples of the alarming environment and species statistics.

*"75% of the planet's land surface has been "significantly altered." (2news)*

*"66%of the ocean area is experiencing multiple impacts from people, including from fisheries, pollution, and chemical changes from acidification" (119website)*

*"Our climate is warming at an alarming and unprecedented rate" (3wiki)*

*"Nearly 72% of global greenhouse gas emissions are generated by just 15 countries." (4website)*

*"29 percent of the Great Barrier Reef coral has died in 2016." (5movie) (114website)*

*"50% of the Borneo Rain Forests have been lost in the past 50 years" (6book)*

*"Glaciers have receded by an extraordinary amount in the last few years."* **(7movie)**

> *"Canada is warming twice as fast as the rest of the world, according to a new government report. That report also warned that **drastic action is the only way to avoid catastrophic outcomes**"* (83website).

The Tipping Point could well be reached when a combination of two or more of the following might well become catastrophic.

- Alaska permafrost (8news): release of methane
- Siberian forest fires (9website)
- UK peat bogs (10news)
- Indonesian forest fires (11website)
- Mangrove swamps (12website)
- Reduced snow reflection (13news)
- Antibiotics resistance (14website)
- Communicable diseases and viruses (15website)

## Atmospheric Green House Gas (GHG) emissions

**Business As Usual** cumulative global emissions for 2020 to 2050 are expected to be 2252 Gigatons $CO_2e$. A reduction of 1185 Gigatons is estimated to be required to have a 50% chance of achieving the maximum 2 degree rise in the earth's temperature.

The vast majority of greenhouse gas emissions come from a small set of countries. Their source is predominantly energy use, such as electrical power plants, industrial processes (cement, aluminium and steel manufacturing), vehicles, and buildings. ***Figure 5.1*** shows the relative country contributors.

Figure 5.1: Global Green House Gas (GHG) emissions by sector

*Green house gas emissions split by sector is in the order of:*

| | |
|---|---|
| *Energy supply (incl electricity and heat):* | *26%* |
| *Industry* | *20%* |
| *Agriculture:* | *14%* |
| *Forestry* | *17%* |
| *Transport* | *13%* |
| *Residential and commercial* | *8%* |
| *Other* | *2%* |

Energy supply and Industry account for almost half of the emissions.

The electricity sector is responsible for the largest share of emissions due to the combustion of coal, gas, oil, and biomass for power generation.

Industrial process emissions are dominated by releases related to natural gas and petroleum systems, as well as cement, steel and aluminium production.

Land use includes livestock and landfills.

An overview of Key Process Industries is shown in **Figure 5.2.**

## Figure 5.2: Key Process Industries and GHG Emissions

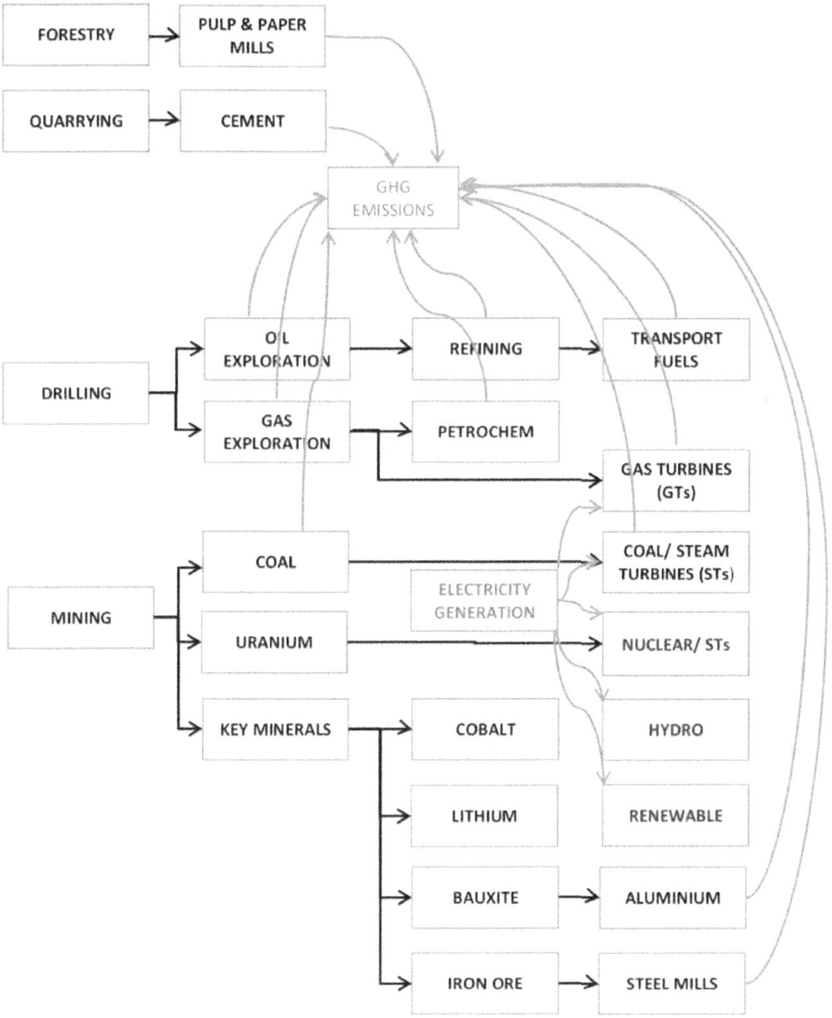

Emission reduction potential (20website) could be split as follows:

1. Industry Efficiency Standards              16 %
2. Industry Process Emissions Policies        10%
3. Complementary Power Sector Policies        11%
4. Renewable Portfolio Standards              10%

| | |
|---|---|
| 5. Carbon Pricing | 26% |
| 6. Land Use | 15% |
| 7. Building Codes and Appliance Standards | 5% |
| 8. Fuel Economy Standards | 3% |
| 9. Urban Mobility | 2% |
| 10. Vehicle Electrification | 1% |
| 11. Other | 1% |

As we can see from this list, the largest emission abatement potential (73%) comes from energy use and industrial processes.

## Methane (17website)

*Anthropogenic effects, processes, objects, or materials are those that are derived from human activities, as opposed to those occurring in natural environments without human influences.*

Methane is the second most abundant anthropogenic green house gas (GHG) after carbon dioxide ($CO_2$), accounting for about 20% of global emissions. Methane has the ability to trap heat in the atmosphere 20 to 30 times greater than $CO_2$. As a result, methane emissions have contributed to about one-third of today's anthropogenic GHG warming.

### Oil and gas production

Air emissions and waste water generation are the immediate negative results of oil and gas exploration with the accompanying high potential for oil spills. (See Exxonmobil, BP and Texaco/Chevron disasters in **Chapter 3**). Energy consumption in exploration and conversion of oil and gas into salable products (16book) is shown as follows:

- Liquefied Natural Gas (LNG):     7-8 %
- Gas To Liquid (GTL):     25 %
- Oil Refining:     5 %

The 'Big Five' International Oil Companies - BP, Shell, ExxonMobil, Chevron and Total - have shown little movement towards reducing their fossil fuel base until recently. Shareholder pressure has forced Shell, BP and Total to established plans to reach net-zero emissions by 2050. ExxonMobil, Chevron have not committed to anything other than near term reduction of climate warming pollution (132news).

Burning of low grade gas by flaring is a major part of oil and gas production. To reduce flaring, Maersk Oil Qatar was seeking Carbon Credits for its oil well production unit, but the required procedures were not followed and the credits were not approved. **Box 5.1** outlines the project.

### Box 5.1: Flaring elimination project (16book)

*A Production Sharing Agreement (PSA) contractor implemented a flare reduction project on its offshore oil platform. Regrettably, the planned Carbon Emission Reductions (CERs) were not obtained as the proper process was not followed for registration/accreditation of a clean development mechanism project with the Clean Development Mechanism (CDM) Executive Board of the UNFCCC.*

Rasgas and Qatargas joined forces to eliminate Boil Off Gas (BOG) when loading Liquefied Natural Gas (LNG) tankers. **Box 5.2** outlines this project.

### Box 5.2: Capex justification—environmental compliance and Return on Investment (RoI) (16book)

*In the past 15 years, the LNG industry has grown exponentially with significant advances in technology. The flaring of boil-off gas (BOG) when loading LNG tankers has become a major environmental issue due to the massive increase in export of LNG. Rasgas and Qatargas, two large integrated gas plants in Qatar, joined forces to invest in a BOG recovery system. The scale of the project made it partially economically viable, with the recycling of the BOG to the process units. It is anticipated that carbon credits for the project may also be traded in the future, thus making it even more financially feasible.*

## Energy consumption

Companies, including State Operated Enterprises (SOEs), such as electric power generators, oil refineries, petrochemical plants, steel mills, cement plants, paper mills, etc., are the biggest contributors towards global warming.

### Electrical power generation

Both State Owned Enterprises (SOEs), which produce and distribute electric power, as well as private power generation companies, tend to use the cheapest energy resource for the generation of power, which, in many countries, tends to be coal. Coal is the dirtiest resource available but is also the most carbon-intensive fossil fuel we can burn. In **Box 5.3** I give an example of poor judgement in the issuing of a loan to promote sustainable power generation.

### Box 5.3: World Bank Loan (18news)

*Eskom is a SOE which has a monopoly on electric power generation in South Africa. In April 2010 the World Bank Group's International Bank for Reconstruction and Development approved a US$3.75-billion loan to Eskom, justified under the guise of promoting renewable and clean energy. Despite this rhetoric, $3.05-billion of the loan was dedicated to completing the Medupi coal-fired power station in Limpopo. This power station will be the largest coal-fired power plant on Earth, emitting more carbon dioxide than the 143 least-emitting countries in the world.*

*Only $260-million was dedicated to building the Sere Wind Farm and the Upington Concentrated-Solar Project.*

The most polluting electricity-producing power plants use coal, while gas power plants are much less polluting and more efficient. Electricity generation from nuclear, hydro, wind and solar does not emit any gases into the air and so some electric power generating companies are diversifying their generation to include renewables such as solar and wind.

## Process industries consumption

Cement and steel plants are energy intensive and have to use the cheapest energy source available, and this is often (dirty) coal. Steel mills have a high particulate level in the emissions causing serious air pollution local to the steel works. Hebei Province in China is an example (80website). Pulp and paper mill emissions tend to be from waste forest products that are used to generate power.

Gas plants typically feed process plants which produce plastics and chemicals. The waste gas is often used for the processes and power generation. Qatar Portland Cement Company (QPCC) uses clean low grade waste gas from the gas plants in the country.

Aluminium appears to be a good construction material as it is lightweight and doesn't rust and it is also 100% recyclable. Unfortunately, aluminium smelters are electrical power intensive and tend to be situated where there is plenty of cheap power, such as low grade gas as in Qatar, or hydro power as in Norway or Canada. **Box 5.4** outlines the the devastating effect of establishing large dams for hydro power for aluminium smelters, resulting in the elimination of large parts of rainforest and the displacement of indigenous people.

### Box 5.4: The Dam Problems in Sarawak, Borneo Malaysia

*Bakun Dam and the Aluminium Smelter (19website)*
*Sarawak Report* 16 February 2012
*'**Scandal of SALCO - How Taib Plans To Make Billions From Bakun.**
**Exclusive!***

*Displaced families, who have received so little compensation from Bakun, may be interested to learn how Taib and his own family are planning to make billions for themselves out of the hydro-electric dam, based on an investment of just RM2.00! We can reveal that SALCO (Sarawak Aluminium Company Sdn Bhd), the company that has been handed a licence to build a vast smelter in Simalaju, powered by cheap electricity from the dam, is secretively entirely owned by the Taib family company, CMS.'*

*(see also **Box 3.4**)*

**Murum Dam and resettlement** *(85website)*
BRUNO MANSER FUND, BASEL / SWITZERLAND 07 August 2014
**'Malaysian lawyers slam Sarawak dam resettlements**
*The Malaysian Bar Council, the official association of all Malaysian advocates and solicitors, slams Sarawak Energy, a state-owned Malaysian power supplier, for the resettlement conditions of 1500 indigenous Penan and Kenyah who were displaced by the recently-constructed Murum dam. According to a report published today, the controversial dam in the Malaysian state of Sarawak in Borneo has led to "deplorable" living conditions and a "decimated" livelihood for the natives.'*

## Forestry and farming

## Logging

Logging and palm oil production (21news) have been the most devastating on the world's rain forests. The rain forests are logged, leaving poor soil. Palm oil is then planted and, being a monoculture, saps the soil of energy which makes it difficult for anything else to grow.

## Agriculture

Tilling (plowing) in preparation for planting crops causes the soil to release vast amounts of $CO_2$ into the atmosphere (128movie).

The use of chemical fertilizers and pesticides is reducing the soil's ability to rejuvenate. In addition to this, the effect of pesticides on humans is now being revealed resulting in lawsuits against manufacturers such as Bayer, the maker of Roundup (129news).

## Genetically Modified (GM) seeds and foods (115website)

Some GM seeds are patented and have to be purchased each season. The advantage is that they produce larger, more pest resistant crops, but don't necessarily guarantee good crops. Farmers using GM seed stock run the risk of falling into debt traps since they can't harvest and grow their own seed. GM foods have helped feed millions but self sustaining subsistence farming has suffered in the process.

## Cattle farming (22news)

Overgrazing of cattle is contributing to desertification (128movie) and the methane generated by cattle is a major issue estimated to be up to 27% of global anthropogenic methane emissions (17website).

An entrepreneur in South Africa is trying to address this in a small way by providing manure from his cows to a biogas plant (116video). 25,000 cows provide 40,000 tons of cow manure and this natural resource together with food waste produces 4.4 Mw of 100% 'green' electricity.

### Plastic pollution (23news)(24news)

According to the World Bank, humans generated 242 million tons of plastic waste in 2016. Statistics from 2015 show that about 55% of plastic waste was dumped, 25% incinerated and 20% recycled. Current action to reduce plastic pollution is discussed in the paragraph *"Leading change - actions by companies"*. I given an example of action being taken in **Box 2.6**.

# Evidence of global warming and deterioration of Earth

There appears to be a great deal of controversy around global warming, with many refusing to believe that the phenomenon even exists. Evidence is essential to convince and influence change. I have listed some key documentaries, conferences and reports which provide irrefutable proof of this crisis.

## Documentaries, talks and websites

This is a list of excellent awareness documentary films, videos, talks and websites which aim to increase awareness of issues affecting the sustainability of businesses of the future.

I have identified a few popular films documenting the degradation of Earth.

### An Inconvenient Truth 2006 (26movie)

Al Gore explains how humans have destroyed the planet. Gore issues an urgent warning on what must be done to save the Earth.

### An Inconvenient Sequel: Truth to Power 2017 (118movie)

Al Gore continues his tireless fight, travelling around the world to train an army of activists and influence international climate policy.

### Chasing Ice 2012 (7movie)

The loss of glaciers around the world over time is documented.

### Chasing Coral 2017 (5movie)

The bleaching of large parts of the Great Barrier Reef in just a few years is shown.

### A Plastic Wave 2018 (130video)

This film documents just how serious the problem of plastic pollution is.

There are a number of Ted Talks raising awareness of environmental issues. I recommend a select few.

In 'How to restore a rainforest' (61tedtalk) Willie Smits describes how he believes he has found a way to re-grow clearcut rainforest in Borneo by piecing together a complex ecological puzzle, saving local orangutans and creating a thrilling blueprint for restoring fragile ecosystems.

In his talk 'Why we're storing billions of seeds' Jonathan Drori (124tedtalk) encourages us to save biodiversity - one seed at a time. Reminding us that plants support human life, he shares the vision of the Millennium Seed Bank, which has stored over 3 billion seeds to date from dwindling yet essential plant species.

In 'The other inconvenient truth' Jonathan Foley (125tedtalk) shows why we desperately need to begin "terraculture" - farming for the whole planet. A skyrocketing demand for food means that agriculture has become the largest driver of climate change, biodiversity loss and environmental destruction.

Websites are also used productively to promote relevant issues. One that I particularly like, Mongabay (86news)(127website), gives updates on the latest conservation news from around the world. Sadly, Mongabay's reporting of environmental corruption in Indonesia has possibly led to the detention of their editor in January 2020.

There has been a backlash from major companies who have vested interests and have used scientific advisers to cast doubt on what climate change experts have been saying over the years (25book)(117movie). This is very disturbing to any person interested in saving the planet.

## Conferences

### 21st Conference of the Parties (COP21) of the UN Framework Convention on Climate Change (UNFCCC): Paris (27wiki)

The Paris Agreement is an agreement within the United Nations Framework Convention on Climate Change (UNFCCC), dealing with greenhouse-gas-emissions mitigation, adaptation, and finance, signed in 2016. As of February 2019, 194 states and the European Union have signed the Agreement. 185 states and the EU, representing more than 87% of global greenhouse gas emissions, have ratified or acceded to the Agreement, including China, the United States and India, the countries with three of the four largest greenhouse gas emissions of the UNFCCC members total (about 42% together). On August 4, 2017, **the Trump**

Administration delivered an official notice to the United Nations that the U.S. intends to withdraw from the Paris Agreement **(28website)** as soon as it is legally eligible to do so. **However, President elect Joseph Biden plans to reverse this decision. The USA is the 2<sup>nd</sup> biggest emitter of GHG.**

## 2016 United Nations Climate Change Conference: Marrakesh (29wiki)

This was an international meeting of political leaders and activists to discuss environmental issues. The conference incorporated the twenty-second Conference of the Parties (COP22), the twelfth meeting of the parties for the Kyoto Protocol (CMP12), and the first meeting of the parties for the Paris Agreement (CMA1).

Four important conclusions came from the Marrakesh gatherings:

1. **Our climate is warming at an alarming and unprecedented rate,**
2. We have an urgent duty to respond,
3. Sustainable development, climate and humanitarian needs are interconnected and are urgent priorities for our world and
4. These challenges cannot be solved in isolation.

## G20 Nations (30news)

G20 countries are outlined in *Chapter 2 paragraph "Supra National Groups"*.

> *"Japan wants to make reducing the glut of plastic waste in the oceans a priority at the Group of 20 summit it is hosting in June 2019 as governments around the world crack down on such pollution."*

The EU has voted to outlaw 10 single-use plastic items, including straws, forks and knives by 2021. It has also set targets for all plastic packaging, the top source of plastic waste, to be recyclable by 2030.

Asia is also home to the biggest contributors to the estimated 8 million metric tons of plastic that winds up in the ocean each year (31news). The top five marine plastic trash offenders are all in Asia: China and Indonesia - both G-20 members - followed by the Philippines, Vietnam and Sri Lanka.

## 2019 UN Conference for Action (CAS 2019) (32website)

*"Climate change is the defining issue of our time and now is the defining moment to do something about it. There is still time to tackle climate change, but it will require an unprecedented effort from all sectors of society."*

The Summit delivered a major step up in national ambition and private sector action on the pathway to the key **2020 climate deadline**.

The UN Secretary General demands to world leaders and CEOs at the Climate Action Summit were:

1. No new coal: No new funding or construction of coal facilities from 2020

2. No fossil fuel subsidies: Stop spending $4.7 trillion per year on fossil fuel subsidies

3. Net zero in 2050: Commit to carbon neutrality ie net zero emissions by 2050

4. Make polluters pay: Tax polluters and cut taxes for people.

The closing press release is shown in ***Box 5.5***.

### Box 5.5: CAS 2019 Closing Press Release (edited)(32website)

*65 countries and major sub-national economies such as California committed to cut greenhouse gas emissions to net zero by 2050, while 70 countries announced they will either boost their national action plans by 2020 or have started the process of doing so. Over 100 business leaders delivered concrete actions to align with the Paris Agreement targets, and speed up the transition from the grey to green economy, including asset-owners holding over $2 trillion in assets and leading companies with* *combined value also over $2 trillion. Many countries and over 100 cities - including many of the world's largest - announced significant and concrete new steps to combat the climate crisis. Many smaller countries, including Small Island Developing States and Least Developed Countries, were among those who made the biggest pledges, despite the fact the they have contributed the least to the problem.*

*Among the major announcements today:*

• *France announced that it would not enter into any trade agreement with countries that have policies counter to the Paris Agreement.*

• *Germany committed to carbon neutrality by 2050*

• *12 countries today made financial commitments to the Green Climate Fund, the official financial mechanism to assist developing countries in adaptation and mitigation practices to counter climate change….*

• *The United Kingdom today made a major additional contribution, doubling its overall international climate finance to 11.6 billion pounds for the period from 2020 to 2025*

• *India pledged to increase renewable energy capacity to 175GW by 2022 and committed to further increasing to 450GW…*

• *China said it would pursue a path of high quality growth and low-carbon development….*

• *The European Union announced at least 25% of the next EU budget will be devoted to climate related activities.*

• *The Russian Federation announced that they will ratify the Paris Agreement, bringing the total number of countries that have joined the Agreement to 187.*

• *Pakistan said it would plant more than 10 billion trees over the next five years.*

*On unprecedented levels of private sector action:*

• *A group of the world's largest asset-owners -- responsible for directing more than $2 trillion in investments -- committed to move to carbon-neutral investment portfolios by 2050.*

• *87 major companies with a combined market capitalization of over US$ 2.3 trillion pledged to reduce emissions and align their businesses ... to limit the worst impacts of climate change—a 1.5°C future.*

• *130 banks – one-third of the global banking sector – signed up to align their businesses with the Paris agreement goals on transitioning from brown to green energy:*

• *Michael Bloomberg will increase the funding and geographic spread of his coal phase out efforts to 30 countries. Already, his work has helped to close 297 out of 530 coal plants in the US.*

• *Countries, including France and New Zealand, announced that they will not allow oil or gas exploration on their lands or off-shore waters.*

• *Heads of State from Finland, Germany, Greece, Hungary, Ireland, Italy, Netherlands, Portugal, and Slovakia, are among those that announced that they will work to phase out coal. South Korea announced it would shut down four coal-fired power plants, and six more will be closed by 2022…*

- *The Summit also delivered critical platforms for improving energy efficiency and reducing the growing energy needs for cooling….*
- *Further, the Climate Investment Platform was officially announced today. It will seek to directly mobilize US$ 1 trillion in clean energy investment by 2025 in 20 Least Developed Countries in its first year.*

*The Secretary-General urged governments, businesses and people everywhere to join the initiatives announced at the Summit….*

## Conference of the Parties COP 25 2019 (33website)

COP25 Climate Change Conference took place in Spain in December 2019 where countries and other organizations committed to addressing global warming.

**Reports**

## UN Intergovernmental Panel on Climate Change (IPCC) Fifth Assessment Report (AR5) (34news)(35website)

The report states that **urgent and unprecedented changes are needed to reach the target**, which it says is affordable and feasible although it lies at the most ambitious end of the Paris Agreement pledge to keep the global warming to between 1.5C and 2C.

## UN State of Earth Report 2019 (36news)(37website)(38news)

This is the IPBES Global Assessment Report on Biodiversity and Ecosystem Services published by UN Intergovernmental Science-Policy Platform on Biodiversity and Ecosystem Services (UN IPBES). It was approved by the IPBES Plenary at its 7th session in May 2019 in Paris, France (IPBES-7). A brief summary of the report follows.

- **75% of the planet's land surface has been "significantly altered."** Biodiversity "is declining faster than at any time in human history."
- **"One million species already face extinction**, many within decades, unless action is taken to reduce the intensity of drivers of biodiversity loss. Without such action there will be a further acceleration in the global rate of species extinction, which is

already at least tens to hundreds of times higher than it has averaged over the past 10 million years."

- **The global biomass of wild mammals has declined by 82%** since prehistory. It's not just the number of species that's crashing, it's the total amount of creatures, too. One way to measure this is by mass.

USA Fourth National Climate Assessment (NCA4) (39wiki)

The authors say that **without more significant mitigation efforts, there will be:**

> **"substantial damages on the US economy, human health, and the environment.** *Under scenarios with high emissions and limited or no adaptation, annual losses in some sectors are estimated to grow to hundreds of billions of dollars by the end of the century."*

While the Climate Science Special Report (CSSR) is "designed to be an authoritative assessment of the science of climate change" in the US, it does not include policy recommendations. *Box 5.6* gives details.

The Trump Government response to this report was that it is to be redacted so as not to cause alarm (40audio).

### Box 5.6: US Government Climate Assessment (39wiki)

*Fourth National Climate Assessment (NCA4) 2017/2018 is a two-part congressionally mandated report by the us Global Change Research Program (USGCRP)... The report, which took two years to complete, is the fourth in a series of National Climate Assessments (NCA) which included NCA1 (2000), NCA2 (2009), and NCA3 (2014).*

*Like the previous reports in this series, the NCA4 is a "stand-alone report of the state of science relating to climate change and its physical impacts".*

*Volume 1 "Climate Science Special Report" (CSSR) was released in October 2017. Researchers reported that "it is extremely likely that human activities, especially emissions of greenhouse gases, are the dominant cause of the observed warming since the mid-20th century. For the warming over the last* century, there is no convincing alternative explanation supported by the extent of the observational evidence."

Volume 2 "Impacts, Risks, and Adaptation in the United States", was released on November 23, 2018. According to NOAA, "human health and safety" and American "quality of life" is "increasingly vulnerable to the impacts of climate change".

## BP Statistical Review of World Energy 2019 68th Edition (41website)

*"...events of 2018: a year in which there was a **growing divide between societal demands for an accelerated transition to a low carbon energy system and the actual pace of progress**"*

*"In particular, the data ... in 2018, global energy demand and carbon emissions from energy use grew at their fastest rate since 2010/11, **moving even further away from the accelerated transition envisaged by the Paris climate goals**"*

The report is summarized as follows.
- Global primary energy consumption grew rapidly in 2018, led by natural gas and renewables. Nevertheless, carbon emissions rose at their highest rate for seven years.
- **Primary energy consumption per capita** by country is an interesting new statistic where primary energy comprises commercially-traded fuels, including modern renewables used to generate electricity.

# Action being taken

Action being taken is manifested through policy decisions by countries, influence of groups, economics and business initiatives.

## Outcomes of the UN Climate Action Summit 2019 (COP 25)

*"Climate science is clear: the world faces a massive ecological and humanitarian crisis. The climate emergency is the defining and most*

*urgent issue of our time, and it cannot be avoided without a global shift away from fossil-fuel dependency." (33website)*

Commitment, action, proof of action and impact on climate change was required to be reported on with the initial report back at the Madrid conference in December 2019. The expected outcomes were not realised. Observers of the summit held G20 countries (especially the US, Brazil, Australia, Saudi Arabia) and major oil, gas and coal companies responsible for undermining the climate ambition and blocking the progress for better response to this global challenge (42website).

## United Nations Summit on Biodiversity September 2020

Political leaders participating in the United Nations Summit on Biodiversity in September 2020, representing 64 countries from all regions and the European Union, have committed to reversing biodiversity loss by 2030 (120website).

## Government policies to drive change (43website)

Government policies will drive change to a carbon neutral environment. The key policies are:

> In the power sector, renewable portfolio standards and feed in tariffs can reduce emissions by increasing the share of fossil-free power generation. Complementary power sector policies such as support for transmission and smart distribution management are also required.

> Standards and incentives to improve industrial energy efficiency can significantly reduce emissions from the industrial sector. Compliance with standards, such as ISO 14001 Environmental management systems, can assist with self regulation and improvement.

> In the building sector, building codes and appliance standards are the best tools for reducing emissions.

> Fuel economy standards, electric vehicle incentives and smart urban

transport planning can reduce emissions significantly in the transportation sector.

Carbon pricing is another strong tool for reducing emissions, encouraging emission-reducing behavior across the economy and pushing investments to lower-carbon options.

European member countries are in the forefront of establishing policies to move towards carbon neutrality by 2050.

## Economics of change

A continual reduction in the cost of sustainable energy is driving the move from fossil fuel generation. Wind and solar cost of generation is now competitive with fossil fuel generation with a lower investment cost. Nevertheless the wind does not blow all the time and the sun is only available less than half the time.

The pressure to reduce a company's carbon footprint is increasing, yet energy resource companies are making large profits and so are reluctant to change. Some of the biggest companies in the world (104website) are from this sector:

- ExxonMobil
- British Petroleum
- Shell
- Sinopec
- China National Petroleum
- Glencore

For all that, the reduction of even an energy's company carbon footprint is possible (see **Box 6.8**). Finnish petrochemical company Neste (44news) is in the forefront of innovative change in line with the Paris Accord.

> *"Today, more than 70% of Neste's profit comes from the area of renewable products, which are produced from waste and residue."*

Economics of change is evident in various countries as indicated by the following.

The US has officially entered the 'coal cost crossover' (45website), where existing power generated from coal is increasingly more expensive than cleaner alternatives. Today, local wind and solar could replace approximately 74% of the coal fired power stations at an immediate savings to customers. By 2025, this number is predicted to grow to 86 % of the coal stations.

Since 2005, Europe's power industry (46news) has had to pay for the majority of its $CO_2$ emissions and this has triggered a change in profit calculations. Thus, coal faces stiffer competition from cleaner-burning natural gas.

**Leading change**

Climate change groups

Climate change groups are essential in influencing policy makers and I have listed some here.

Nearly 180 companies have signed the Science Based Targets initiative, which sets emissions reduction targets. 69 companies have joined the RE100 initiative (see *later paragraph* ), committing to 100% renewable power, and many are more than halfway to meeting this target.

The Green Climate Fund (87website) is a fund established within the framework of the UNFCCC to assist developing countries in adaptation and mitigation practices to counter climate change.

Renewable energy groups include International Renewable Energy Agency (IRENA), Renewable Energy and Energy Efficiency Partnership (REEEP), International Energy Agency (IEA), Renewable Energy Policy Network for the 21st Century (REN21), and International Solar Alliance (ISA) (88website).

The Climate Group's Mission (47website) is "Accelerating climate action" and its goal is "A world of no more than 1.5°C of global warming and greater prosperity for all". The Climate Group brings together powerful networks of businesses and governments, acting as a catalyst to focus on the greatest global opportunities for change, shifting global markets and policies towards its goal.

The Carbon Disclosure Project (CDP) (133website) is an international, nonprofit organization providing the only global system for companies and cities to measure, disclose, manage and share vital environmental information. CDP holds the largest collection of primary climate change, water and forest risk commodities information and puts these insights at the heart of strategic business, investment and policy decisions.

We Mean Business is a global coalition of nonprofit organizations working with the world's most influential businesses to take action on climate change. Its collective mission is to ensure that the world economy is on track to avoid dangerous climate change by 2020, while delivering sustainable growth and prosperity for all. The coalition brings together seven organizations:

1. BSR
2. **CDP**
3. Ceres
4. The B Team
5. **The Climate Group**
6. The Prince of Wales's Corporate Leaders Group
7. World Business Council for Sustainable Development.

RE100 (48website) is led by The Climate Group in partnership with CDP, as part of the We Mean Business coalition and represents the world's most influential companies committed to 100% renewable power. RE100's purpose is to accelerate change towards zero carbon grids on a global scale. Switching corporate energy demand to renewables is transforming the global energy market and accelerating the transition toward a clean economy. Companies include Ikea, HP, Google, Apple, GM, Tata.

Oil and Gas Climate Change Initiative (OGCI) (51website) is a voluntary CEO-led initiative taking practical actions on climate change. OGCI members leverage their collective strength to lower carbon footprints of energy, industry, transportation value chains via engagements, policies, investments and deployment. Representation is from the big oil and gas companies, including British Petroleum, Royal Dutch Shell, Chevron, ExxonMobil, Total, Saudi Aramco, China National Petroleum Company, ENI, Petrobras, Pemex, Repsol, Occidental and Equinor.

The Alliance to End Plastic Waste (AEPW) (49website)(50news) was formed in January 2019. Its mission is to develop, deploy, and bring to scale, solutions that will minimize and manage plastic waste and promote solutions for used plastics. This includes plastic reuse, recovery, and recycling to keep it out of the environment. Its projects focus on river renewal, municipal and city partnerships, idea incubation and IT infrastructure. Members include Chinese state-owned Sinopec, one of the largest companies in the world. The Alliance has committed to a goal of $1.5 billion to deliver sustainable solutions over five years.

The Global Methane Initiative (GMI) (84website), launched in 2004, is an international public-private initiative that advances cost-effective, near-term methane abatement and recovery and use of methane as a clean energy source in three sectors:

1. oil and gas systems

2. coal mines

3. biogas (including agriculture waste, municipal solid waste, and waste water).

Working in collaboration with other international organizations, the Initiative has formed key alliances with partners such as the United Nations Economic Commission for Europe (UNECE) and the Climate and Clean Air coalition (CCAC) to reduce global methane emissions.

## Countries actioning change

In addition to the commitments of the UN Conference for Action 2019 listed in **Box 5.5**, the following are specific examples of what some countries are doing.

The USA is a leading in some areas yet is lagging in other areas. In spite of the Trump Administration of the US Federal Government withdrawing from the Paris Accord, various states are moving ahead quickly to establish more sustainable power projects. Solar power in California has been growing rapidly as a result of community support, declining solar costs, and a Renewable Portfolio Standard which requires that 33% of California's electricity comes from renewable resources by 2020, and 50% by 2030 (89wiki). A Renewable Portfolio Standard set by the Arizona Corporation Commission requires 15% renewable energy by 2025 among regulated utilities, 4.5% of which must come from distributed renewable energy sources (90wiki). In December 2015 Nevada (91movie)(92wiki) had restrictions placed on home installation of solar cells which were then lifted after a public outcry.

Brazil is the world's second largest producer of ethanol fuel (93wiki). Brazil's 40-year-old ethanol fuel program is based on the most efficient agricultural technology for sugarcane cultivation in the world, using modern equipment and cheap sugar cane as feedstock. The residual cane-waste (bagasse) is used to produce heat and power. Ethanol fuel consumption in the country achieved a 50% market share of the gasoline-powered fleet in February 2008. In terms of energy equivalent, sugarcane ethanol represented 17.6% of the country's total energy consumption by the transport sector in 2008. However, there is a down side to this: **the Amazon rain forest is being destroyed to create more sugar plantations** (see *Chapter 6 paragraph "Biofuels, sustainability and ethics"*).

The European Commission (EC) is initiating the European Green Deal investment plan (105news), also known as the Sustainable Europe investment plan. The EC will devote more than half-a-trillion euros from the EU budget over the period 2021 to 2030 to the plan, and will leverage an equivalent amount from other public and private sources. The Green Deal will create *"an enabling framework for private investors and the public sector to facilitate sustainable investments and provide support to public administrations and project promoters in identifying, structuring and executing sustainable projects"*.

Denmark's long term sustainable energy policy has permitted companies developing alternative energy technologies to take a long term perspective. Denmark is now a top wind turbine manufacturer, producing the largest wind power turbines in the world.

Regional electricity power production (52website) is taking off in Africa. The World Bank has a number of projects aimed at generating off grid power in Central Africa.

South Africa (54news)(55wiki)(56video)(94news) is proceeding rapidly in developing power generating plants using wind and solar. It plans to allocate at least 6 GW to large scale solar by 2030. This is equivalent to about 3 times the power requirements of Cape Town. Photovoltaics would supply 10.5% and wind power is expected to make up 22.5% of the energy mix. Albeit coal will still form around 43% of South Africa's generation capacity.

In Asia, Bhutan (57tedtalk) is currently Carbon Neutral and will proceed to being Carbon Negative by 2030. **China, the biggest contributor to global warming, has now committed to being carbon neutral by 2060 (126website).**

Australia (58news) is moving rapidly towards renewable energy sources. In Adelaide, Elon Musk has developed a battery which has proven very successful for storing energy (60news).

> "The Australian renewable energy pipeline is surging above 100GW of solar, wind and utility storage projects with investment matching upstream capex at $10bn per annum," (59website)

India plans to have 175 GW of renewables by 2022 (see **Box 5.5**).

In Japan, recent nuclear disasters have initiated a major shift towards electricity production from gas turbines.

China's contribution to global renewables growth (45%) is more than the entire OECD combined (41website).

Indonesia (61tedtalk) now has a comprehensive forest certification program for export of hard woods (see **Box 1.16**).

## Actions by companies

**Corporate Policy Change** (62website) has been initiated by a major sector of US business. In August 2019, the US Business Round Table, representing 180 plus CEOs of US companies, changed its policy to be more inclusive with respect to stakeholders other than shareholders (see **Box 1.9**).

Investors (68news)(95news) are increasingly focused on the threat of global warming and the need to rapidly decarbonise the global economy, with up to $118tn of funds committed to making climate risk disclosures by 2020.

Public sector organizations, faith-based groups, foundations, and universities (64website) have shown their support for curbing emissions by withdrawing their investments from fossil fuels. In total, more than 550 institutions with assets of $3.4 trillion (a much smaller portion of which is invested in fossil fuels) have divested. Britain's biggest asset manager LGIM (65news) removed Exxon from its 5 billion pound ($6.3 billion) Future World funds for what it said was a failure to confront threats posed by climate change. Sarasin and Partners stated in July 2019 that it had sold nearly 20% of its holdings in Shell, saying its spending plans were out of sync with international targets to battle climate change.

In September 2019, nine development banks—The World Bank Group, New Development Bank, the Islamic Development Bank, IDB Group, European Investment Bank, the EBRD, AIIB, African Development Bank Group and ADB—agreed to a five-point plan (106news). It includes "helping our clients deliver on the goals of the Paris Agreement" and "support increased climate finance levels". The plan also states that "each institution will take actions to help clients move away from the use of fossil fuels".

Under pressure from their shareholders, some large oil companies are now beginning to reduce their carbon footprints. British Petroleum CEO announced in February 2020 that it planned to be net carbon neutral by 2050. British Petroleum (BP) Ventures (66news) has made an investment of $30 million in Calysta, an alternative protein producer, that will use BP's natural gas to produce protein for fish, livestock and pet feeds. BP has also agreed to form a joint venture with Bunge (67news), a leader in agriculture and food, that will create a leading bioenergy company in one of the world's largest fast growing markets for biofuels. BP is also part of a consortium to develop the net zero Teeside project (96website)

**Electricity generation companies are increasing renewables in the mix of power generation offered.** Wind and solar generation are the primary sources and I discuss some examples of companies in the renewable power generation field here.

Accoina (69website) is the biggest global energy company operating exclusively in the renewable energy sector, present in over 20 countries on five continents. It works with five technologies: wind, solar photovoltaic, hydroelectricity, biomass and solar thermal.

Mainstream Renewable Power (70website) delivers wind (onshore and offshore) and solar photovoltaic plants across diverse markets. Their business model is to take projects through the development process, de-risk them and sell them at financial close, during construction or at commercial operation. Their vision is to electrify the world with renewable energy.

North Sea Wind Power Hub (NSWPH) (71website) is developing an estimated capacity of 180GW by 2045, to provide clean power to "hundreds of millions of Europeans".

Royal Dutch Shell and Dutch gas company Gasunie (107news) plan to build a massive green hydrogen plant in the northern Netherlands in the next 10 years to be fueled by a large new wind farm off the Dutch coast The plant will ultimately be able to produce 800,000 tons of hydrogen by 2040. (1 ton of hydrogen is equivalent to about 1000 gallons of gasoline.)

**Transport fuel companies are researching alternatives to fossil fuels.**

A $5 billion dollar green hydrogen production facility is to be sited in Neom (112news), Saudi Arabia. The project will use about 4 GW of renewable power from solar and wind to produce 650 tons of carbon-free hydrogen per day. The green hydrogen will be used to power buses and trucks globally and will be equally owned by partners Air Products, ACWA Power, and Neom.

Solar Fuels Research (72video) has achieved major breakthroughs in the production of ethanol, methane and hydrogen from the sun. Ethanol could be used for blending with petrol, methanol for space heating, while hydrogen can be used to power fuel cells. Fuel cells are already being used on Vancouver buses (97website).

Biofuels have affected the environment both positively and negatively. One source, palm oil, is the result of stripping rainforests to plant palm oil trees.

Total (73news)(74news) converted it's La Mede refinery to produce biofuels. It plans to produce both biodiesel and biojet fuel for the aviation industry, and it will also produce premium hydrotreated vegetable oil (HVO), known as renewable diesel. Total has said it will produce the biofuels using around 60% to 70% sustainable vegetable oils including rapeseed, **palm oil** and sunflower oil, and 30% to 40% from treated waste from animal fats, cooking oil and residues. However, French legislation (98news) removed palm oil from a list of permitted biofuels in January 2020 which will prohibit the La Mede refinery from using it. The European Union also plans to restrict the use of palm oil in biofuel due to the environmental impact, something which has triggered diplomatic tensions with top producers Malaysia and Indonesia (see also **Box 1.13**).

> *"Neste's target is to become a global leader in renewable and circular solutions..."* Lars Peter Lindfors, Senior Vice President, Innovation at Neste.

Neste (75news) is the world's leading provider of sustainable renewable diesel, renewable aviation fuel, and an expert in delivering drop-in renewable chemical solutions. Neste MY Renewable Diesel is available in Finland, Estonia, Latvia, Lithuania and Sweden, as well as in California and Oregon in the United States (99news). It is made from 100% renewable raw materials and emits up to 90% less greenhouse gas emissions compared to conventional fossil diesel. A large variety of waste fats are used in the production, including waste and residue from the meat, fish and vegetable oil industries. Neste MY Renewable Diesel has the same chemical composition as fossil diesel and easily replaces conventional diesel or can be blended in any proportion. The company has recently signed a long-term agreement for wind power with a leading clean-energy company Fortum.

In the USA, Diamond Green Diesel (108news), a renewable diesel, is being produced by the joint venture of Valero Energy and Darling Ingredients, using the Honeywell Ecofining process developed in conjunction with Eni of Italy.

The difference between biodiesel and sustainable diesel is described in **Box 5.7**.

*Box 5.7: Bio diesel vs renewable diesel*

*Bio diesel is recommended for mixing with fossil diesel normally up to a maximum of 20%. It is generally not used on its own as it could cause engine trouble. The main concern is that bio diesel is less stable than fossil based diesel, and deteriorates over time. Light, temperature and humidity can all increase the rate of deterioration. Another concern is that bio diesel is often derived from palm oil.*

*On the other hand, renewable diesel can be used as a 100% substitute for fossil diesel as it has the same chemical composition.*

Lufthansa started using Neste's sustainable aviation fuel (100news), blended with fossil jet fuel on flights departing from Frankfurt in 2019. Neste's sustainable aviation fuel is produced from renewable waste and residue raw materials. Over the lifecycle, including for the impact of logistics, sustainable aviation fuel has up to 80% smaller carbon footprint compared to fossil jet fuel. With the Singapore refinery expansion on the way, Neste will have the capacity to produce over 1 million tons of renewable jet fuel by 2022. The company is also investigating production of liquid fuels from plastic waste.

SkyNRG sustainable Aviation Fuel (76news), the Amsterdam-based global market leader for sustainable aviation fuel (SAF), is building a facility at Delfzijl in the Netherlands. SAF will be supplied directly to aircraft at the nearby airport, and as a strategic partner, KLM Royal Dutch Airlines has committed to buy 75,000 tons of SAF per year for a ten-year period. SAF reduces $CO_2$ emissions by at least 85% compared to conventional jet fuel. The 100,000 tons produced at the facility will reduce $CO_2$ emissions from aviation by more than 250,000 tons.

LanzaTech (111website), a leading US biotech company and carbon recycler, has successfully launched LanzaJet, a new company that will produce sustainable aviation fuel (SAF).

German utility company, EWE (101news), is planning pilot projects for supplying hydrogen to meet the growing demand for cleaner fuels and potentially add a new area of business by the middle of the decade. A surplus of onshore and offshore wind can be turned into hydrogen through electrolysis and then be stored and transported like natural gas.

## The Chemical industry is also undergoing change towards sustainability

Clariant, a world leader in specialty chemicals, has signed an agreement for a new partnership with Neste (77news). The raw materials, C2/C3 monomers, are derived from Neste's renewable hydrocarbons produced from 100% renewable feedstock. Through the partnership, sustainable polyolefin solutions derived from renewable hydrocarbons can be offered. Neste has also entered into a strategic co-operation agreement with Borealis (102news) for the production of renewable polypropylene (PP). The co-operation will enable Borealis to start using Neste's 100% renewable propane as renewable feedstock at its facilities in Belgium.

Since plastics are such major pollutants, a lot of thought is going into the design of products to minimize the use of plastics and to ensure maximum recycling of the packaging and the product at the end of its life. Volkswagen (122website) who use large amounts of plastics during the manufacture of their cars, consider the entire life cycle of each component during the design phase.

Plastics recycling has gained greater emphasis in recent years. Plastic recycling (103website)(123website) refers to the process of recovering waste or scrap plastic and reprocessing the materials into functional and useful products. There are six common types of plastics as follows:

1. PS (Polystyrene)

*Example: foam hot drink cups, plastic cutlery, containers, and yogurt.*

2. PP (Polypropylene)

*Example: lunch boxes, take-out food containers, ice cream containers.*

3. LDPE (Low-density polyethylene)

*Example: garbage bins and bags.*

**4. PVC** (Polyvinyl chloride)

*Example: cordial, juice or squeeze bottles.*

**5.HDPE** (High-density polyethylene)

*Example: shampoo containers or milk bottles.*

**6.PET** (Polyethylene terephthalate)
*Example: fruit juice and soft drink bottles.*

Currently, only **PET, HDPE, and PVC** plastic products are easily recycled, although plastics that cannot be recycled can be converted into fuel and petroleum-based products. PET is recycled to produce clothing and thermal insulation among a host of other products.

Dow Chemicals in partnership with UPM Biofuels (109news), a producer of advanced biofuels, announced the commercialization of plastics for the packaging industry made from a bio-based renewable feedstock which originates from sustainability managed forests. The entire supply chain is International Sustainability & Carbon Certification (ISCC) certified, based on mass balance approach, meaning all steps meet traceability criteria and reduce negative environmental impacts.

Spanish company Repsol (110news) is the first petrochemical company to certify all its complexes for the production of circular polyolefins. This is in line with International Sustainability & Carbon Certification (ISCC). Feed oil is obtained from plastic waste which is not suitable for mechanical recycling.

**$CO_2$ conversion and disposal technology will have a major impact on reducing global warming.** Oil and gas companies are already reinjecting $CO_2$ into old oil and gas wells to enhance production.

Carbon recycling firm LanzaTech (79news)(80website), a Chicago-based startup, has cut multi-million-dollar deals on four continents, including in China, aimed at helping curb climate change and air pollution. At the Jingtang Steel Mill in Hebei province, LanzaTech uses bacteria to convert waste gases into ethanol for use in cleaner fuel, capturing carbon so that it is not released into the atmosphere as planet-warming $CO_2$.

ExxonMobil and Global Thermostat (121website) have signed a joint development agreement to advance breakthrough technology that can capture and concentrate carbon dioxide emissions from industrial sources, including power plants, and the atmosphere. The company (78news) has also said it will invest up to $100 million over 10 years to research and develop advanced lower-emissions technologies. In my view, this investment is a drop in the ocean for the biggest company in the world: annual investment of 10 million is 0.05% of 2018 earnings.

An innovative Indian Company, Graviky Labs, (81news) has invented technology to remove carbon from exhausts to produce printer ink, called Air-Ink.

With respect to **conserving the rainforests**, Willie Smits (61tedtalk) has not only created a rainforest, but even changed the circumambient climate. Companies, such as Kimberly Clark, are producing paper products from sustainable forests (see ***Box 4.5***).

**Commercial fish farming** has grown considerably. Vietnamese fish farmed in the sheltered waters of Ha Long Bay near Hanoi are found in the frozen aisles of South African supermarkets, on the opposite side of the world. This has a large carbon footprint impact considering the carbon emissions from airfreight from Vietnam to South Africa. However, an enterprising farmer at the southern tip of Africa is farming trout in a local dam and tanks. ***Box 5.8*** describes a three party agreement for water use.

### Box 5.8: Baardskeerdersbos Trout Farm

*Water is a controlled resource in South Africa, where a permit has to be obtained from the government to dam rivers or dig boreholes.*

*A vineyard owner wanted to establish a dam on a river flowing through his farm. However, to obtain a permit to do so, he had to agree to support the local municipality and community by supplying water to the municipality and allowing trout farming in the dam.*

*Besides the vineyard having a plentiful supply of water, the local municipality now has an additional source of water and a local trout farmer raises fingerlings in the dam for her trout enterprise.*

Abalone poaching off the southern coast of South Africa was once very profitable since the mollusk fetches high prices in China, where it is regarded as a delicacy. However, it was notoriously difficult to police. The introduction of commercial farming of abalone has virtually stopped the poaching and dozens of abalone farms have popped up along the coast, thereby yielding legal earnings for many enterprising businesses.

## Conclusion

Global warming and deterioration of the state of Earth are very real, as supported by overwhelming evidence. Governments and companies are starting to look towards cleaner alternatives due to the influence of both organisations as well as people on the ground, in spite of initial resistance 2020 is the year to get major projects off the ground so that results can be seen by 2030. Sustainable businesses that take initiatives to become carbon neutral by 2050 will thrive and those persisting with 'business as usual' will disappear.

# CHAPTER 6: THE FUTURE - WHAT CAN BE DONE

## Introduction

In this chapter I present alternative scenarios based on what has happened to companies in the past 30 years. Stakeholder dynamics of the past are not necessarily an indication of how they will play out in the future, where changing and more volatile influences will be imposed on companies.

I express my predictions of the types of companies that will thrive and those that are likely to succumb.

## The next 30 years

### Overview

**Good governance and the application of sustainability principles** are going to be key driving forces in changing to a carbon neutral world. So who is going to be around in 2050? Developing countries with growing populations will be hotbeds for entrepreneurs. In China, the progressive opening up of the economy and allowing grass roots democracy is critical to their success and others in the region. Countries that are adopting more inclusive governance codes will be the ones to watch, for instance Germany, Japan and New Zealand.

**Transnational businesses** will become more responsive to stakeholder pressure to comply with global warming and other environmental issues. Human rights issues are also being highlighted with a reduction in the use of indentured labour. Pressure is being put on companies that source goods from countries that have questionable human rights records. Shareholders will have more say in determining the direction of these big companies. Royal Dutch Shell shareholders are challenging management to be more focused on energy savings as well as reducing the impact Shell has on the environment. Top management bonuses are accordingly related to improvement in these fields. Other large oil companies are moving to becoming 'energy' companies with a greater focus on renewable energy.

**Countries** such as Denmark are also setting higher targets for energy from renewable sources. In 2018, Scotland had a 24-hour period of just renewable energy supply for the first time. France is committed to being carbon neutral by 2050 and the small Himalayan country Bhutan is already carbon neutral. Developing countries do not necessarily have the 'baggage' of existing fossil fuel infrastructure that rich countries have and can therefore leapfrog to renewable energy more easily. With the advent of cellphones and off-grid power generation, more remote communities are not dependent on their government for the supply of telecoms and power.

**Group influence** on governments and companies is going to be more intense. In the US, the well oiled lobbying process, which has been dictated by big business to maximize profit, must swing more towards social and environmental issues. The alternative will be increased unrest among the poor as witnessed after the Wall Street Crash of 89.

**Small businesses**, from poor to rich communities, will flourish by taking advantage of technology. Payment methods are greatly simplified with the use of cell phones. Technological innovations, such as on-line platforms, have already enabled Indian farmers to cut out middle men (98 website)(99website).

En-route to 2050, the UN has set 2030 as the year to **eliminate extreme poverty** (1). India and China are making great progress towards this goal.

## Timeline

**The next ten years will be critical for reversing the loss of biodiversity and determine whether we can achieve a carbon neutral world by 2050.** The change from 'business as usual' to a sustainability focused business requires a change in behaviour from all members of society.

I have proposed key milestones in **Table 6.1**.

## Table 6.1: Milestones to a Carbon Neutral World

| ACTIONERS | 2020 | 2030 | 2040 | 2050 |
|---|---|---|---|---|
| COUNTRIES | Decide national actions and target dates. Start actions. Publish decisions. | Interim targets achieved? Take corrective actions. | Additional measures achieved? Take corrective actions. | Carbon neutral |
| COMPANIES | Decide company strategic objectives and target dates. Start actions. | Interim targets achieved? Take corrective actions. | Additional corrective actions to get on target. | Carbon positive |
| GROUPS | Decide critical focus areas for influencing politicians and boards and apply influence. | Influence politicians and boards to take corrective actions. | Continue the campaigns | Keep the pressure on defaulters |
| INDIVIDUALS | Establish personal carbon footprint. Influence all around you. | Influence all around you. | Influence all around you. | Carbon positive footprint |
| All | Work towards UN Sustainable Development Goals set in 2015 | Achieve goal of eliminating extreme poverty | | |
| MILESTONES | COP 25 2019: Global warming Action plans agreed. UN Summit 2020: Commit to reverse biodiversity loss by 2030. Pursue UN goal set in 2013 to end extreme poverty. | Max 1.5°C global warming - the turning point. Biodiversity loss reversed. Extreme poverty eliminated. | Positive tipping point. | Achieve a carbon neutral world |

To assist in the process of behavioral change, Ted Talks has launched Countdown (91tedtalk), a global initiative to champion and accelerate solutions to the climate crisis, by turning ideas into action. Countdown's goal is to build a better future by cutting greenhouse gas emissions in half by 2030 in the race to a zero-carbon world . Every organization, company, city, nation and their citizens are invited to collaborate with Countdown and take action on climate change.

## Applying scenarios

As discussed in **Chapter 1**, scenarios are powerful tools which can be used to find alternative outcomes for the future. Shell has been a leader in this field (7video) (8website).

Scenario planning was used to prepare for the transition from apartheid to democracy in South Africa, so as to achieve a peaceful change (5book) (6website). **Box 6.1** outlines the example. The transition to a carbon neutral future requires scenario planning on a national level so as to debate the options open to countries and be able to determine appropriate action.

Box 6.1: From Apartheid to Democracy

*Up until 1994, South Africa was controlled by white supremacists who were forced to make a transition to a real democracy. Fear of outright civil war and insurrection was evident. Scenario planning was used to explain the options open to South Africans. The South African scenarios: the high road and the low road How will the transition go and will the country succeed in taking off? The options focused on the choices facing the country as to whether, through consultation and negotiation, it did what was necessary to travel on the 'high road' to a non-racial democracy and rising prosperity, or whether it continued as a repressive, centralized society and controlled economy which would continue on the 'low road' of confrontation, conflict and falling incomes leading inexorably to a 'waste land'. The scenarios were presented to some 230 audiences totaling 25 000 to 30 000 people. The audiences varied widely: the Cabinet, government departments, 'homeland' governments, political parties and clubs and associations across South Africa. On April 27, 1994, millions of South Africans voted in the country's first fully democratic elections. Just two weeks later, Nelson Mandela was sworn in as president, in a ceremony that filled people with promise. He was the first black South African president, chosen in free and fair elections.*

*The first element of the 'high road' was well and truly evident.*

## Slowing global warming

**Global warming is going to be one of the most significant influences on the world of tomorrow. The transition to reducing global warming is simply not happening fast enough** (2website).

> "..in 2018, global energy demand and carbon emissions from energy use grew at their fastest rate since 2010/11, **moving even further away from the accelerated transition envisaged by the Paris climate goals.**"

**Chapter 5** discusses global warming in detail under the topics of effect, evidence and action.

The United Nations report on global warming and the extinction of species is disturbing.

### Headline "Human society under urgent threat from loss of Earth's natural life" (3news)

It has become clear that global warming is going to affect countries, companies, groups and individuals, in short, everyone. We all need to apply ourselves urgently to reduce the rate of increase in global warming. The United Nations Framework Convention on Climate Change (UNFCCC) and others are guiding this process at an international level, but we cannot just leave it to large organizations. Grass roots global warming reduction initiatives (such as Al Gore's free courses for global warming leaders) are urgently needed.

### Global warming scenarios

*Figure 6.1* shows the **best, most likely and worst case scenarios**. As per the 2018 BP Global Energy Review (2website), **business was still continuing as usual.**

## Figure 6.1: Global Warming Scenarios (2website)

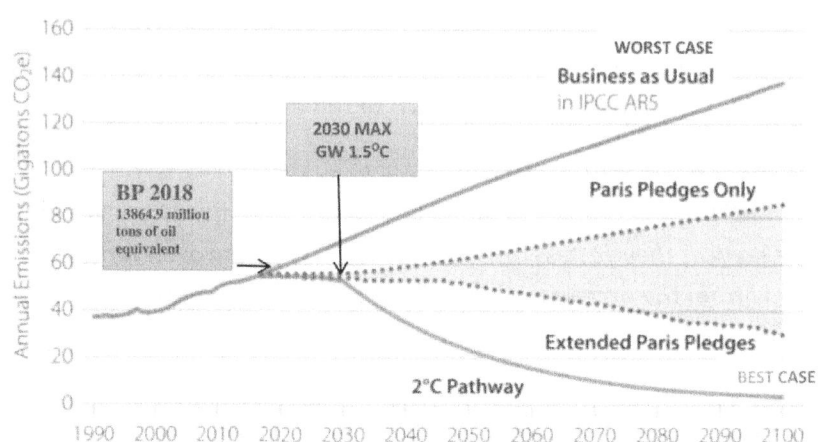

Sustainability initiatives to achieve the 2030 target are discussed in **Chapter 5**.

## Green House Gases (GHGs)

Nearly 75% of global greenhouse gas emissions are generated by just 20 countries. **Figure 6.2** shows the $CO_2e$ emissions by country.

Figure 6.2: $CO_2e$ emissions (4website)

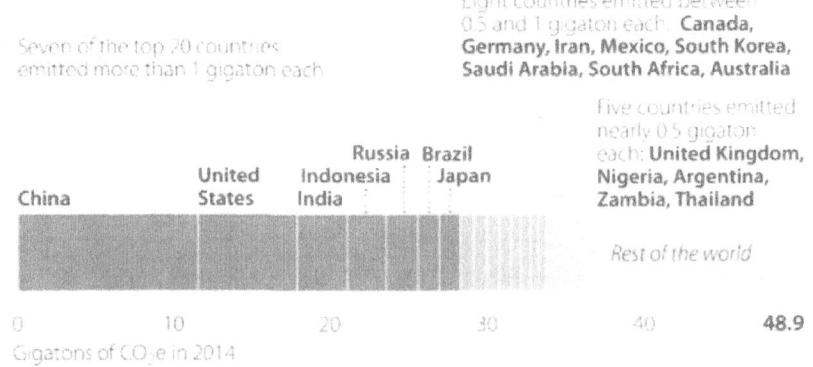

**Behavioral change by leaders in government and industry is critical for change to be effective** and **positive action by the top emitters will make an immediate and substantial difference.** Unfortunately, leaders of G20 countries and major oil, gas & coal companies at the UN Summit in Madrid (COP25) had not been convinced of the urgency of change.

## Addressing water availability and consumption

The ancient Khmer civilization of Angkor was highly dependent on the water system that they created. This collapsed due to climate change, which they, in effect also created, is regarded a major factor in the demise of the Khmer empire (94).

As a result of global warming and the increase in city populations, the availability of clean water to large cities is going to become more critical. In 2014, intense droughts dried up at least six lakes around Chennai. Some areas had water in their pipes only once every three or four days, for just an hour or two. In 2015, São Paulo had less than 20 days of water left when the rains eventually came (100website). In 2017 Cape Town was the first big city to experience a major water shortage where it was predicted to the day that, should there be no rain, water would run out (32website). The disaster was averted by farmers who supplied their excess water to the city, and then shortly afterwards rain alleviated the crisis.

Water for agriculture is continuing to stretch available sources in some areas, for example on farms in the Darling Basin in Australia.

There is some light at the end of this dry tunnel though. Reverse osmosis is becoming more cost effective to produce clean water. Abu Dhabi is planning to use this technology to desalinate seawater in one of the largest, and most cost and energy efficient plants in the world. The plant is scheduled to commence full commercial operations in 2022 (29news).

## Energy production and consumption: greening and curtailment

Energy consumption is the greatest contributor to global warming and emissions, with fossil fueled power plants and process plants being the main culprits. The reduction in the use of fossil fuels and the capture of $CO_2$ from process plants will go a long way towards curtailing the increase in global warming.

Current reduction Initiatives:

- Investment in carbon capture technology by ExxonMobil (10news) and others
- Institutional investors moving away from investment in fossil fuel industries towards renewables (11news)
- Production of bio fuels and chemicals but **not** from mono crops such as palm oil and sugar cane which are destroying virgin forests (13news)
- Use of electric vehicles as these become more competitive with petrol and diesel vehicles (14website) (15news)
- Changing demand from high carbon footprint consumables to low carbon footprint consumables (16website)
- Power generation from renewables and gas in preference to coal (12website).

Power generation from gas will continue to become a more economic and cleaner option than that from coal. The increased supply of cheap gas and the higher thermal efficiency of Combined Cycle Gas Turbine (CCGT) electricity generators (30news) are major factors promoting the change to power from gas. Typical thermal efficiency for electrical generators is around 37% for coal compared to 56 – 60% for combined-cycle gas-fired plants.

Wind and solar power are also becoming more economically viable, although the storage and conversion of this energy remains problematic. Excess wind power could be stored in massive batteries or converted to hydrogen. Shell in the Netherlands is planning a massive green hydrogen production facility using power from offshore wind farms (31news).

## What can the top global warming contributors do?

The top 20 were identified in **Box 4.1**. What can these and other companies do, to ensure a carbon neutral economy by 2050?

The basic principles of sustainability are:

1. To do no harm through our activities

2. To replace what we have taken

3. Where it is not possible to replace, then substitute something of equal value to our community and the physical environment.

It probably is not possible for these companies to comply with the first two objectives, but the third is certainly attainable. For example, by diversifying investment into carbon capture technology, sustainable fuels & petrochemicals, commercial forestry, renewable energy etc.

An analogy may help by comparing two industries to highlight an industry in decline vs an industry that could prevent a similar decline by transitioning to being carbon neutral. Box 6.2 illustrates this.

> Box 6.2: Demise of the tobacco industry vs renewal of the oil & gas industry
>
> *Various lawsuits have been brought against the tobacco industry in the US (18website) and Canada (19website). In spite of this, Philip Morris, the biggest tobacco manufacturer, still makes huge profits. At least, it is no longer in the top 25 companies in the world. This is illustrated below.*

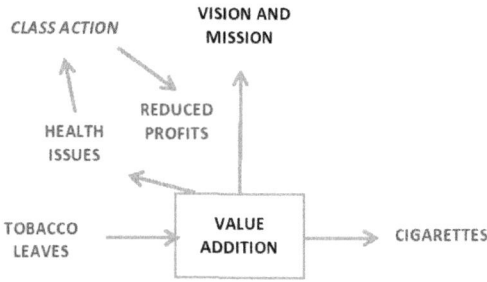

The big oil and gas companies need to take heed and provide greater commitments to becoming carbon neutral by 2050. It is possible as we can see in the following figure.

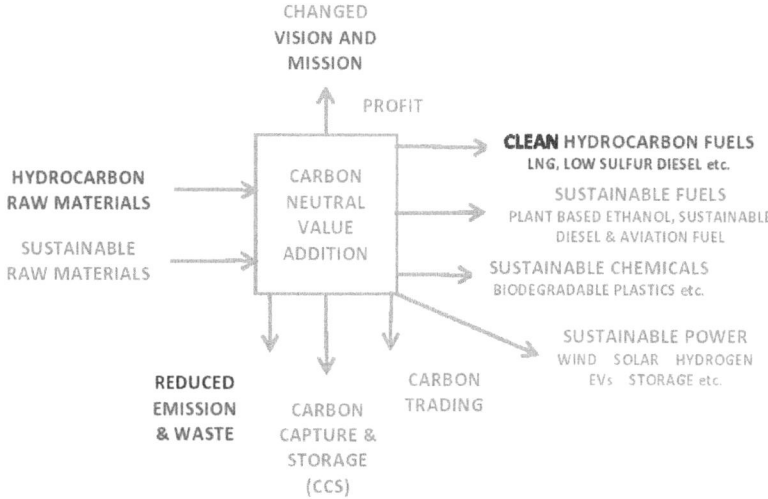

Neste, an oil refining and marketing company in Finland, is already obtaining more than 70% of its profits from renewables, even using wind power (34news) at its production sites.

Europe's top oil firms have all set carbon reduction goals: BP has set Scope 3 targets, Repsol aims to become a net zero-carbon company by 2050, Shell has set out an "ambition" to halve Scope 3 emissions by 2050 and Total has set a goal to cut Scope 1 & 2 greenhouse gas emissions. Scope 3 is explained in **Box 6.3**.

Box 6.3: GHG emissions - scope (35website)

*Greenhouse gas emissions are categorised into three groups or 'scopes' by the most widely-used international accounting tool, the Greenhouse Gas (GHG) Protocol.*

*Scope 1 covers direct emissions from owned or controlled sources. Scope 2 covers indirect emissions from the generation of purchased electricity, steam, heating and cooling consumed by the reporting company. Scope 3 includes all other indirect emissions that occur in a company's value chain.*

Singapore LNG importer Pavilion (36news) is planning to implement a standardised industry framework for LNG GHG emissions and pave the way towards more environmentally responsible and sustainable natural gas strategies. These strategies include increased long-term demand in carbon offsets as the key to unlocking additional GHG mitigation projects that benefit communities and the environment. Such projects could include forest conservation or the development of significant renewable power generation that would not have otherwise occurred.

BP and US-based independent Dominion Energy announced in February 2020 that they have joined the ranks of energy companies adopting a long-term net-zero emissions target. Other companies to have announced net-zero, or near net-zero goals include Spain's Repsol, Norway's Equinor, Sweden's Lundin Petroleum and London-listed Energean Oil & Gas. All have set their target for 2050 apart from Lundin, which is outdoing them all by aiming for carbon neutrality by 2030 (37news).

US oil majors have lagged behind their European rivals, with Chevron setting limited reduction targets and ExxonMobil having no targets at all.

## Bio fuels, sustainability and ethics

*"On Wednesday, November 6, 2019, Brazilian President Jair Bolsonaro and his Ministers of Economy and Agriculture jointly signed a decree that extinguishes the environmental zoning for*

*sugarcane that had restrained the advance of this crop into the Amazon and the Pantanal (Brazil's famous wildlife-rich wetlands)."* (88website)

Many bio fuels have been found to be detrimental to the environment and may contribute to global warming. The stripping of the Amazon rainforest for sugarcane plantations to produce bio ethanol , the logging of the Borneo rainforests to plant oil palms for the production of bio diesel (see **Box 1.13**) and the wholesale logging of US forests to feed wood chip power plants (83website)(84movie) are seriously detrimental to our environment. Greater awareness of the disadvantages of some bio fuels is necessary and policymakers should be advised accordingly. People who believe that all bio fuels are beneficial to the environment need to be informed and influence the politicians who are making the decisions. **Ethical leadership is desperately needed.**

# Replicating success

Replication of successful programs can rapidly increase the rate of change. Individuals, groups, companies and countries need to be able to replicate success formulas. Here are a few examples of replication.

### Corporate governance (King IV) (41website)

Concepts that form the foundation stones of King IV are: ethical leadership, the organization in society, corporate citizenship, sustainable development, stakeholder inclusivity, integrated thinking and integrated reporting. This South African Corporate Governance Code is one of the most enlightened codes available and, ideally, should be applied in all developing countries.

### Corporate Social Responsibility (CSR)

CSR programs that are working in one community should be replicated in other communities.

The rapid decline in the stock of affordable housing in Seattle (42news) has resulted in a large homeless population. In 2019 Microsoft unveiled its $500 million fund, which earmarks $475 million for investments in affordable housing projects and $25 million in philanthropic grants to organizations tackling homelessness. The launch of Microsoft's initiative kicked off a series of similar housing commitments from other tech companies in the months that followed. Facebook and Google launched housing programs of their own, each valued at about $1 billion. Amazon and Salesforce made similar commitments. Apple became one of the biggest spenders on housing, launching a 2.5 billion dollar effort to alleviate California's housing crisis.

Goedgedacht (43website) started in the South African winelands 25 years ago with the motto 'putting rural children first'. They clothe, feed and educate rural children in their villages and they have created a Path Onto Prosperity (POP) model **which is being replicated around the country**. They have also created 9 youth centres. Goedgedacht is supported by the biggest industry in the area: Pretoria Portland Cement.

## Credit and financial transactions for small businesses (micro lending)

Small businesses in poor communities often need small amounts of credit and also need to be able to transact simply without having to go, in person, to the nearest bank.

Muhammad Yunus established a microfinance organisation and community development bank, Grameen Bank, in Bangladesh. He received the Nobel Prize for his efforts in 2006. In February 2011, Yunus co-founded Yunus Social Business (YSB) Global Initiatives (39website) which creates and empowers social businesses, which address and solve social problems around the world.

> "We unite the power of women with the power of technology to increase women's access to essential services. And we help to build an environment that supports her breakthrough, whether through

*strengthening peer networks or developing products designed for her."* YSB Quote

In East Africa, mobile banking (44news) has leapfrogged conventional banking and its online counterpart for everyone, from the rich to the very poor, at an astonishing pace. More than 18 million customers now use mobile phones to perform banking transactions in Kenya and Tanzania. A whole new generation of customers' very first bank account is accessed purely through their phones. So-called 'mobile banks' are springing up worldwide as a growing number of people join the cashless economy. Since its launch in 2007 in East Africa, the M-Pesa service (M stands for mobile and pesa means money in Swahili), launched by mobile network operator Safaricom, has been phenomenally successful. This banking concept has now spread to the rest of the developing world.

## Environment

> *"The twin perils brought by climate change, an increase in the temperature of the ocean and its acidity, if they continue to rise at the present rate the reefs will be gone within decades and that would be a global catastrophe."*
> Sir David Attenborough, Natural Historian and British Broadcaster (45website)

Coral farming has become critical as coral reefs are disappearing at an extraordinary rate. The process involves collecting coral from reefs and growing it in a nursery until it has fully matured, before returning it to the ocean at restoration sites. Coral reef farming, also known as coral aquaculture, is quickly being acknowledged as the best solution for rebuilding the habitat of many unique species as well as reducing the pressure on the reefs and reef organisms that overfishing has caused.

Off the east coast of Malaysia, a team is attempting to rebuild coral reefs (46website) by rescuing broken coral fragments from surrounding reefs and re-planting them onto specifically designed concrete blocks called 'Reef Squares'. The team aims to re-transplant 2500 coral fragments a year. This effort is being replicated elsewhere and many thousands of coral fragments are being re-introduced into the ocean.

The Green Fins (47website) initiative aims to protect and conserve coral reefs through environmentally friendly guidelines that promote a sustainable diving and snorkelling industry. Green Fins has a proven conservation management approach which leads to a measurable reduction in negative environmental impacts associated with diving and snorkelling. The approach is replicable, and has been adopted by 11 countries and nearly 600 individual marine tourism companies since its inception in 2004.

Sea turtles play a vital role in ocean ecosystems (48website). According to the International Union for Conservation of Nature (IUCN), many turtle species are listed as threatened, endangered or critically endangered. Population decline is attributed to a long history of egg exploitation, commercial hunting and harvesting of marine turtles, fishing mortality, loss of nesting habitats, marine pollution, negative impacts of tourism and the lack of a national strategy on marine turtle conservation.

Private turtle conservation groups are being replicated around the world, such as the Lang Tengah Turtle Watch (LTTW) (46website) who have teamed up with a 5 star resort. The resort has benefited by an increase in their occupancy rates as a result of their association with the LTTW conservation program. (see **Box 6.9**)

Rain forests are being destroyed at a phenomenal rate. Some countries simply ignore the issue, for example Sarawak (part of the Malaysian Federation) and Brazil. Others are making attempts to combat corruption and illegal logging such as Indonesia, who have implemented a program to export licensed legal timber (see **Box 1.16**).

Indonesian Willie Smits has proven that the rainforest can be restored (50tedtalk). Hopefully this can be replicated in other rainforests.

## Carbon neutrality

Carbon neutrality within countries needs to be promoted. The small Himalayan country of Bhutan is already carbon neutral (51tedtalk) and hopes to be carbon negative by 2050.

Under its energy plans, Denmark (91website) was Europe's first country to bring in large subsidies for its wind industry, including the feed-in-tariff system, which was successfully replicated in Germany.

***Chapter 7 paragraph "Top Countries and Companies"*** lists countries that have committed to be carbon neutral by 2050.

The annual publication of a company's Carbon Footprint gives incentive for reduction to carbon neutrality. An excellent example is South African retailer Woolworths who publishes its carbon footprint each year (see ***Box 4.3***). Clearly this should be promoted and replicated. Even energy company BP has committed to be carbon neutral by 2050 (52).

## Community projects

The establishment of community projects to spur economic development can be replicated. An example is given in ***Box 6.4.***

> Box 6.4: The Philippi Economic Development Initiative (PEDI)
>
> (92website)
>
> *Philippi has been an agricultural area in the suburbs of Cape Town South Africa since German settlers arrived there over a hundred years ago. They became well-known for their ability to grow vegetables in the sandy soils of the Cape Flats. Today Philippi provides at least half of Cape Town's vegetables but is under threat from the encroachment of industrial and residential properties.*
>
> *The Philippi Economic Development Initiative (PEDI) was established in 1998 as a 'not for profit company' by the City of Cape Town in partnership with the Western Cape Provincial Government, businesses and the community. PEDI's role is to promote and facilitate the acceleration of economic growth. Over a billion rands worth of projects are currently in progress, which will provide*

*opportunities to rebuild and grow this area which has become a shining example of the transformation of disadvantaged communities in South Africa into thriving economic hubs.*

## Using technology

Green technology and bio technology need to take a lead in the sustainability challenge.

Technology has made great strides in many areas such as food, transport, communication, energy, medicine etc., and best of all, has helped to identify how we are destroying our planet so that we can take action to prevent a catastrophic end to civilization on Earth.

Companies are emerging based on inventions utilizing the latest developments in technology. Huawei from China is a leading mobile phone producer and a forerunner in the 5th generation phone technology. Hyundai from Korea is a top car producer, after only starting in the 1960s.

Google Deep Mind AI research (20movie) has replicated a top Go board game champion using neural computing techniques. Evidence suggests that humans will improve by being challenged by AI. Google has also created 'The Moonshot Factory' (21website) as a 'skunkworks' for new ideas.

> *"CCUS is a necessary bridge between the reality of today's energy system and the increasingly urgent need to reduce emissions."*

**Research into carbon capture utilization and storage (CCUS) is increasing.**

A variety of incentives are on offer for further development in this field. For example, the $20M NRG COSIA Carbon XPRIZE (59website) promotes development of breakthrough technologies to convert $CO_2$ emissions into usable products.

New technology for carbon capture is already being employed in a Chinese steel mill (22news).

A demonstration CCUS 50 megawatt power generation unit has been commissioned in Texas USA (53news). It is based on the Allam Cycle: a novel natural gas power plant design that can theoretically capture 100 percent of emissions while being cost- and efficiency-competitive with advanced natural gas plants that have no carbon capture capability. The cycle captures emissions and diverts a pure $CO_2$ output stream into a pipeline for sale or storage, while avoiding most or all water costs and using a fraction of the space of standard natural gas plants.

A consortium, led by Swedish company Liquid Wind (54news), is planning to establish commercial-scale renewable fuel facilities. The Consortium (consisting of Axpo, COWI, Carbon Clean Solutions, Haldor Topsoe, Nel Hydrogen and Siemens) will combine their expertise and technology to produce liquid, carbon neutral fuel from captured carbon dioxide (CCU) and green hydrogen (from renewable electricity).

**Developments in production of green hydrogen** include a Solid Oxide Electrolysis Cell (SOEC ) which is essentially the corresponding Solid Oxide Fuel Cell (SOFC) run in 'reverse' (38website).

Various carbon capture utilization and storage paths are shown in *Figure 6.3*

## Figure 6.3: Carbon capture utilization and storage pathways (78website)

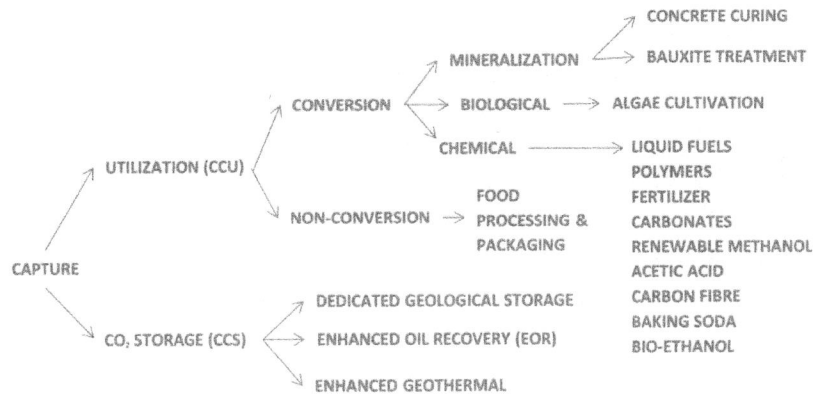

Solar and wind power is required to be used immediately but is not always available. The energy is thus required to be stored for later use. Batteries are used for this purpose, but have limited capacity and are expensive. Molten salts are being used to give limited extension of heat retention for concentrated solar generation plants. Research into using excess solar and wind energy to generate hydrogen is being undertaken (55website). The hydrogen can then be used in power sources such as fuel cells.

Lhyfe (56website) in France is now producing green hydrogen from offshore wind farms using water electrolysis. The difference between Green and Blue hydrogen is explained in *Box 6.5*.

### Box 6.5: Blue vs Green Hydrogen (57news)

**Blue hydrogen** *takes conventional production, largely reforming of natural gas, and adds another step—capture and storage of co-produced $CO_2$ (CCS).*

**Green hydrogen** *dispenses with hydrocarbon feedstocks and splits hydrogen from water, utilising the process of electrolysis. For the hydrogen to be green, the electricity used in this process must be from renewable sources.*

Blue hydrogen production tends to focus on industrial clusters or so-called hydrogen hubs; HyNet in northwest England and H-Vision around Rotterdam are examples. On the other hand green hydrogen production tends to use containerized electrolysers stacked together in a modular fashion on-site and can be set up very quickly.

The NortH2 proposal in the Netherlands envisages that by 2030 up to 4,000MW of dedicated offshore wind could provide annual hydrogen production around four times that of a large-scale gas reforming plant today. By the time blue hydrogen clusters become operational, green hydrogen could be just as cheap and its projects just as large.

Heavy road trucks are being transformed from diesel to electric and hydrogen cell vehicles. Nikola Motor Company (58website) in the USA is an example. The company offers both pure electric as well as hydrogen electric power trains to cover large trucks with Gross Vehicle Weight Rating (GVWR) in excess of 15 tons.

Korean car manufacturer Hyundai (89news) is partnering with Swiss companies to build a value chain covering the production of green hydrogen from hydropower, hydrogen charging stations and the service and maintenance of the trucks. In 2020 Hyundai exported 50 hydrogen-powered trucks to Switzerland.

## Automation, relocation and unemployment

Redundancy is inevitable due to automation as is the outsourcing of manufacturing to those countries that have cheap labour and low tax rates. Those communities which are supported by single large industries experience severe upheaval when the manufacturer automates, relocates or goes bankrupt.

> *Honda, the largest employer in Swindon, England, will cease production in July 2021 resulting in a loss of 3 500 jobs. The company maintains that the move is due to global changes in the car industry*

and the need to launch electric vehicles. *(24news)*

"We're seeing unprecedented change in the industry on a global scale. We have to move very swiftly to **electrification of our vehicles** because of demand of our customers and legislation." Honda Quote

GM closed it's plant in Dayton Ohio in 2009 thereby causing mass unemployment in the town. A Chinese automotive glass manufacturer took over the GM plant, re-employed their workers at lower salaries, and then began to automate large parts of its operation. *(25movie)*

A government framework for training and re-training workers, for progression in existing businesses or for re-employment in new positions, has been established in a number of countries (see **Chapter 2 paragraph "Groups offering training for employment"**).

## Shift to a circular economy

The Platform for Accelerating the Circular Economy (PACE) *(76website)* was launched during the 2018 World Economic Forum (WEF) to drive both public-private action and collaboration to accelerate the transition to a circular economy. Project categories include electronics, plastics, food & bio economy, and business models & markets.

At the 2019 World Economic Forum (WEF) *(27website)*, Google and SAP announced Circular Economy 2030, a $400,000 competition to support entrepreneurs who promote sustainable consumption and production.

### Carbon elimination

Fossil Fuel suppliers and energy intensive industries are moving to carbon elimination in their supply chain as can be seen in ***Figure 6.4***. The retrofitting of existing large emitters with Carbon Capture, Utilization and Storage (CCUS) is required.

## Figure 6.4: Carbon elimination program

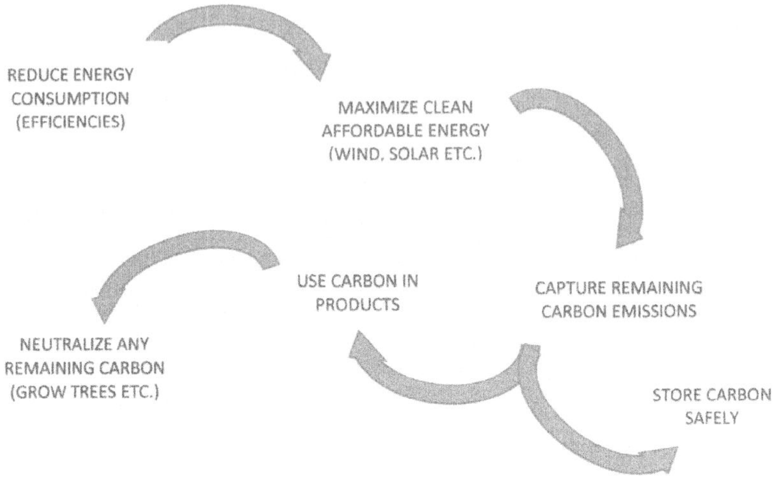

- REDUCE ENERGY CONSUMPTION (EFFICIENCIES)
- MAXIMIZE CLEAN AFFORDABLE ENERGY (WIND, SOLAR ETC.)
- CAPTURE REMAINING CARBON EMISSIONS
- STORE CARBON SAFELY
- USE CARBON IN PRODUCTS
- NEUTRALIZE ANY REMAINING CARBON (GROW TREES ETC.)

Biosequestration is the capture of carbon from the atmosphere by using photosynthesis. A major contribution towards reducing global warming will be through biosequestration, the carbon opportunity created from the rehabilitation of cleared and degraded land (95news)(96website)(97movie).

**Reuse, Renew, Recycle**

Everyone will need to adopt a reuse, renew, recycle lifestyle. Some companies are moving forward rapidly. Volkswagen (40website) already assesses all components of their new cars and trucks, at the design stage, for efficiency of recycling. Some mechanical components from old trucks are currently being returned to VW for re-manufacturing.

It's estimated that by 2050 nearly 60% of plastics demand could be covered by production based on previously used plastics (49website).

# Harnessing the power of the world wide web

The internet has revolutionized our lives, enhancing and creating new industries. Certain restraints and controls are required for the web to be better utilized for the benefit of the global population.

## The Contract for the Web (77website)

This contract was launched in Lisbon at the Web Summit in 2018. ***Box 6.6*** gives an extract of the contract for the web.

> Box 6.6: Contract for the Web extract
>
> "**Governments** must translate laws and regulations for the digital age. ... They must ensure markets remain competitive, innovative and open. And they have a responsibility to protect people's rights and freedoms online...
>
> **Companies** must do more to ensure their pursuit of short-term profit is not at the expense of human rights, democracy, scientific fact or public safety. ... Platforms and products must be designed with privacy, diversity and security in mind....
>
> **Citizens** must hold companies and governments accountable for the commitments they make, and demand that both respect the web as a global community with citizens at its heart....
>
> The fight for the web is one of the most important causes of our time. Today, half of the world is online. It is more urgent than ever to ensure the other half are not left behind offline, and that everyone contributes to a web that drives equality, opportunity and creativity."

## Free access to knowledge

The web has become a source of an infinite amount of knowledge and information. 'Google' has become the word used for searching the web since most of us use Google search engine to find information on just about everything. Wikipedia is a free web based encyclopedia established by Jimmy Wales and Larry Sanger. Kahn Academy (90website) is a nonprofit educational organization with the mission to provide a free, world-class education for anyone, anywhere. Kahn offers school level courses including mathematics, science and reading with an excess of 110 million registered users.

### The power of social media

Social media has become a powerful tool for influencing our shopping, relationships, education and political choices.

All companies can be rated for service and product quality. For example, hotels can be rated by guests. This has become one of the most widely used tools for selection of accommodation.

## What governments can do

Governments need to attract the right companies to create better lives for their citizens. I've listed a few initiatives here.

### Ensure political stability and a favourable investment climate

Political stability is key to enable businesses to operate effectively, but this does not necessarily mean that the political system has to be a western-style democracy.

Key elements of stability include:

   1. Food

   2. Health

   3. Education

An appropriate infrastructure and a staunch anti-corruption stance is required. Examples of what attracts investment include:

   A. A cheap and plentiful supply of electricity, water, educated labour and other resources (for example raw materials)

   B. Being close to the market for their products

   C. An investment friendly government framework with minimum bureaucracy and tax.

*Qatar has the 2$^{nd}$ lowest level of corruption in the Arabian Gulf (33website) and a long term vision to 2030. Milestones toward* achieving this vision include the staging the the Football World Cup in 2022 while their National Oil Company plans to double the output of LNG in the next few years.

China is run by the central committee of the communist party. China's economic growth here been phenomenal with an estimated 850 million people being lifted out of extreme poverty in less than 35 years. That said, neither dissent nor corruption is tolerated, with the death penalty as punishment if government officials are found to be corrupt.

India, the biggest democracy in the world, has excellent telecommunications and rail systems, with technology hubs sprouting in places like Bangalore, but power, water and roads are seriously lacking.

Malaysia has one of the most open economies in the world and offers low cost electricity, water and an educated labour force (60website).

## Promote sustainability

Here are a few key items which responsible governments of the future need to take cognizance of to promote a sustainable future for all.

- Impose a ban on the import of tropical hard woods. Japan is the biggest importer!
- Impose a ban on the import of raw palm oil and bio-diesel derived from palm oil
- Enforce the ban on whaling
- Encourage the application of good governance codes for companies, non profit organizations and government departments.
- Eliminate corruption
- Reduce energy consumption from fossil fuels.

The import of petroleum products is a major source of national expenditure in non-oil producing countries. To encourage reduction in consumption, the UK, Spain, Italy, Germany, France, India and South Africa impose taxes of between 50 and 77% on gasoline and diesel.

Individuals and companies who generate their own sustainable energy should be encouraged, with credits for feeding excess power into the electricity grid.

## Separate executive, judiciary and administrative parts of government

For an effective corruption-free government to survive, each of the primary elements of a government need to be separate. This is required as a balance to check on any excesses of the others, especially the Judiciary checking on the Executive .

In the US, President Trump had a number of judges appointed, for life, to the Supreme Court, effectively moving the Court to the 'white right', while continuing to be unethical in his administration of the country.

In Malaysia, the Executive, under Prime Minister Najib, ran roughshod over the Judiciary. (The leader of the opposition was jailed for sodomy under a weak Judiciary which was controlled by the Prime Minister.) Najib was eventually removed by the electorate.

South Africa has a Constitutional Court which addresses any attempts to breach the Constitution. This Court successfully helped to remove a corrupt President after nearly 9 years.

Recently the UK Prime Minister, Boris Johnson, was overruled by the Judiciary when he tried to suspend parliament.

## Apply a corporate governance code

Governments should encourage all companies, non-profit organizations, state enterprises and government departments to apply the principles of good governance along the lines of King IV.

## Privatize State Operated Enterprises (SOEs)

SOEs are generally inefficient and lack innovation whereas private partnerships can often inject improved efficiency and technology. In a number of countries privatization of power generation and telecoms has been very successful, with the proceeds of the sale of these companies being used for improved infrastructure.

## Establish a vocational training framework

Governments need to establish a framework for the continual training, upskilling and development of the population, with a number of countries having successfully established such schemes. I discuss this in detail in ***Chapter 2 in the paragraph titled 'Groups offering training for employment'***.

## Eliminate corruption

Extensive research shows that the long term reduction of corruption requires the full and active support of civil society and media, as corruption will continue to grow as long as there are willing participants. When a government becomes more intolerant towards corruption, companies will be less inclined to bribe officials.

Governments need to build strong anti-corruption authorities who are able to implement their anti-corruption legislation. A good example is the Western Cape Government of South Africa, run by the Democratic Alliance. Displays in their offices state that corruption will not be tolerated and also clearly specifies those practices which are considered corrupt.

Actions to assist with the elimination of corruption:

1.Adopting ISO 37001 Anti bribery management systems

2.Applying Transparency International (TI) counter bribery business principles

3.Complying with Organization for Economic Cooperation and

Development (OECD) principles

4. Applying the Extractive Industries Transparency Initiative (EITI) standards for natural resources management.

## Eliminate money laundering

Money earned outside an individual's country of tax residence as well as illegally earned revenue can be banked in tax havens. So as to minimize illegal earnings passing through tax havens, governments have attempted to regulate banks, including by ensuring banks request and verify the sources of funds.

Illegal earnings tend to be invested in art and property in countries where the source of funding is not required to be revealed (USA, UK, Dubai etc.). Governments need the cooperation of banks, property agents and art dealers to stem the tide of money laundering.

Whistleblowing has been very effective to combat financial crime such as money laundering. The Panama papers are an unprecedented leak of 11.5m files, and a number of countries have taken legal action against individuals as a consequence (see **Chapter 3 paragraph Whistleblowing**).

Countries also need to introduce effective legislation against money laundering (see **Chapter 2 paragraph Governments)**. UK legislation is an example (see **Chapter 2 paragraph Banks and money laundering)**

## Ensure effective securities oversight (94website)

*A security is any investment that can be readily transferred or sold for cash.*

*Securities regulation comprises the regulation of public issuers of securities, secondary markets, asset management products and market intermediaries.*

*Stocks are a form of security.*

*An exchange is where stocks are traded (example: New York Stock Exchange).*

Investors need to be protected from securities fraud and other unethical practices in the stock markets. Governments, therefore, need to ensure that there is an effective Securities Regulator, such as the US Securities Exchange Commission (SEC), which is backed by laws and regulations that are strictly enforced.

## Avoid debt traps

Governments can be lured into debt traps by countries who offer to build infrastructure projects using low cost loans. South Africa came close to being indebted (and effectively bankrupted in the process), to Russia who offered to build a series of nuclear power stations to overcome South Africa's power shortage (see **Box 4.9**). China has garnered a notorious reputation for offering low cost loans for infrastructure projects to corrupt governments. Corrupt politicians often get removed, leaving a new government to either clean up the mess, or continue pillaging. **Box 6.7** outlines 'China's colonization'

> Box 6.7: China's 'Colonization' (61website)
>
> *Adapted from Human Rights Watch 2019 Report*
> *'China's much-touted "One Belt, One Road" initiative to develop trade infrastructure fostered autocratic mismanagement in other countries. In keeping with Beijing's longstanding practice, Belt and Road loans come with no visible conditions, making Beijing a preferred lender for autocrats. These unscrutinised infusions of cash made it easier for corrupt officials to pad their bank accounts while saddling their people with massive debt in the service of infrastructure projects that in several cases benefit China more than the people of the indebted nation.'*
>
> *Malaysia*
>
> *Malaysian Prime Minister Mahathir canceled three major infrastructure projects financed by Chinese loans amid concerns that his predecessor, Najib Razak, had agreed to unfavorable terms to obtain funds to cover up a*

corruption scandal. However, Mahathir has had to renegotiate two of the contracts as the cancellation costs would have put Malaysia in a worse position with nothing to show for the expenditure to date.

China has, however, built and is operating a complete steel plant on the east coast at Kemaman with little involvement of Malaysians.

### Sri Lanka

Unable to afford its enormous debt burden, Sri Lanka was forced to surrender control of a port to China. The port was built, with Chinese loans but without an economic rationale, in the home district of former President Mahinda Rajapaksa.

### South Africa (62news)

The Chinese are building a self sufficient steel complex in the eastern Limpopo Province of South Africa. It will include its own 4000 MW power plant and will not involve South Africans.

### Kenya

Kenya came to rue a Chinese-funded railroad that offered no promise of economic viability.

Pakistan, Djibouti, Sierra Leone, and the Maldives all expressed regret at having agreed to certain Chinese-funded projects. Talk of a Chinese "debt trap" became common (63book).

## Move to zero carbon footprint by 2050

An awareness of a country's carbon footprint needs to be promoted to encourage citizens to become energy and pollution conscious. It is imperative that all countries become carbon neutral by 2050. (BP Energy Review 2018 (2website) lists the carbon footprints of all countries).

Denmark is a classic example of a country that has addressed the sustainability issue before most other countries. **Box 6.8** outlines the Danish approach.

> ### Box 6.8: Sustainable Energy (91website)
>
> *The Danish government set a target of 50% wind energy in electricity consumption by 2020 as part of its long-term strategy to achieve a 100% renewable energy mix in all sectors by 2050.*
>
> *Continuous government support has been in place since the 1980s, including support to long-term R&D, premium tariffs and the setting of ambitious national targets. All of these have helped the domestic wind industry to expand internationally.*
>
> *Denmark has long been a global centre for the manufacture of wind turbines, with Bonus, LM, Siemens and Vestas, some of the world's leading turbine manufacturing firms.*

Transition subsidies need to be applied prudently to prevent abuse, unlike the US grant system where grants are mostly in the form of tax credits. Half of these tax breaks go to bio-fuels such as ethanol and bio-diesel, mostly from mono crops (82website). Forests are being decimated to feed wood chip power plants (83movie), categorized as bio-fuel power plants.

## Establish disaster quick response and recovery programs

Disasters happen. Learning from history is the best way to establish the most likely cases. Governments need to evaluate likely scenarios and establish quick response strategies for disasters.

Taiwan, Hong Kong, New Zealand and Singapore, who established quick response strategies for epidemics after SARS, were able to respond promptly to contain the Covid-19 virus.

On the other hand, the US government response to Hurricane Katrina was mismanaged and lacked preparation (101wiki). There appears to be a repeat of mismanagement with respect to the Covert-19 epidemic (102news). The Obama Administration set up a pandemic preparedness system which the Trump Administration dismantled (103news).

### Apply the Social Progress Index (SPI) and Social Progress Imperative (64website)(80video)(81book)

Social progress has become an increasingly critical topic for leaders in government, business and civil society. Citizens' demands for better lives are evident in uprisings such as the recent Arab Spring and the emergence of new political movements even in the most prosperous countries. The Social Progress Index is outlined in **Chapter 4 paragraph *"Existing Institutional performance measurements"***. Doughnut Economics is a useful framework for building a new socio-economic vision (see **Chapter 7 paragraph *"Age of adaption"***).

### Streamline infrastructure investment priorities

Governments are having to reorganize State expenditure after the Covid-19 pandemic. It is now even more essential to have a structured approach to prioritizing infrastructure projects to ensure that the money is well spent.

Social cost-benefit analysis (SCBA) provides sound project appraisal and, when systematically applied, a basis for prioritizing projects. To this end, The World Bank has developed the Infrastructure Prioritization Framework (IPF) (75website), a multi-criteria decision support tool, initially piloted in Panama and Vietnam. When large sets of small- to medium-sized projects are proposed, resources for implementation are limited. IPF can be used to inform the selection of projects by combining selection criteria into social-environmental and financial-economic indices.

## What companies need to do

### Ensure ethical leadership and direction

The business environment is changing at an ever increasing pace with the need for ethical leadership and good governance being even more pressing for companies to maintain their focus on a sustainable vision.

Effective management is not enough if managers only excel in a static situation. As part of a company's vision, a commitment to sustainability and social responsibility needs to be demonstrated.

Company profitability should focus on value addition **and** their declared direction statement (vision).

Continual scanning of the technological horizon for new developments is essential as is being prepared for unexpected events.

### Separate Chairman and Managing Director roles

I discuss the rationale behind this in **Chapter 1.** Also see **Box 1.19**. Measurement of progress in this respect is discussed in **Chapter 4**.

### Undertake self assessment

**Business as usual** is no longer an option. Companies need to undertake annual self-assessment through formal strategic planning exercises in order to survive, and thrive. (Self assessment in line with Garratt's 4 levels may be useful (65book)). In this process, the highest levels of ethics need to be inspired and certain critical issues, such as that of sustainability, should be documented for action.

### Seek out social responsibility and sustainable projects that support the company's mission and vision

Social responsibility projects need to be identified and addressed. Microsoft's investment to address the low cost housing shortage in Seattle and the involvement of a local cement company in rural development in South Africa are examples (see **paragraph "Replicating success"**).

Support for environmental projects may actually increase revenue. **Box 6.9** gives an example.

Box 6.9: Turtles and profit (46website)

*The east coast of Malaysia was a major nesting ground for turtles in the 1980s, but poaching and consumption of turtle eggs has decimated the turtle population.*

*The government department of fisheries stepped in by contracting out the beaches to egg collectors to bring the eggs to the government hatcheries for nesting and release. At the time, funds available for purchase of eggs were cut, and so private hatcheries, run by non-profit organizations, were permitted.*

*One such organization teamed up with a 5 star resort to build a hatchery on their beach where guests could sponsor a nest. Each nest owner would receive progress reports of nest checks and a video when their nest of hatchlings was released. The resort has seen increased revenue from guests (new and returning) as a direct result of having a resident hatchery on their beach.*

## Apply the concept of social progress (64website)

There has been a growing expectation that businesses should play a role in delivering improvements in the lives of customers and employees, and protecting the environment, since the financial crisis of 2008. This is known as the social progress imperative.

This movement has catalyzed the formation of local action networks that bring together government, businesses, academia, and civil society throughout the world. These organizations are committed to using the Social Progress Index as a tool to assess strengths and weaknesses, promote constructive dialogue, catalyze change, and improve people's lives.

In Brazil, multinational corporations like Coca-Cola, Natura and Fiat-Chrysler are using customized indices to ensure their supply chains are socially and environmentally sustainable.

### Explore technical innovation

A number of innovative initiatives are being implemented with focus on reducing Earth's carbon emissions.

An example is the Net Zero Teesside project (66website)(84website). It is a Carbon Capture, Utilisation and Storage (CCUS) project, based in Teesside in the North East of England. In partnership with local industry and with committed world class partners, it aims to decarbonize a cluster of carbon-intensive businesses by as early as 2030.

Other examples are mentioned in the paragraph *"Using Technology"*.

### Establish an Enterprise Risk Management (ERM) system

All risks to business ventures need to be evaluated and a system for assessment and treatment of likely risks needs to be established. In addition, should a disaster occur, an Incident Preparedness and Operational (Business) Continuity Management (IPOCM) process is required.

## Actions to achieve a carbon neutral world by 2050

Business activity needs to include a series of projects to cope with the rapidly increasing change. Project Management Institute (85website) uses the term 'Agile Project Management'. There should be a focus on Environment, Social and Governance (ESG) issues and related risks. A proactive approach is essential to stay in business. Actions should:

1. Promote fairness and eliminate corruption
2. Reduce/ eliminate dirty fuels: oil and coal
3. Add $O_2$ to the atmosphere
4. Remove $CO_2$ from the atmosphere
5. Promote renewables: wind, solar etc.

6. Contain nuclear power expansion
7. Stop waste water disposal - recycle
8. Implement CCUS: refineries, power stations steel and cement plants etc.
9. Reduce genetically modified foods
10. Promote seed banks
11. Involve indigenous people: reforestation, fishing, recycling waste
12. Promote carbon footprint awareness: products and people
13. Reduce demand for polluting products
14. Contain palm oil production
15. Change from poppy growing to food growing
16. Stop rain forest logging, mangrove infill, peat bog farming
17. Promote commercial forestry
18. Divest from non-sustainable companies
19. Work to prevent major disasters and corruption scandals.

## Managing change

The world is changing faster than ever with the Covid-19 pandemic giving a step change. It is therefore necessary to have processes in place that will ensure successful outcomes to changes. Kotter's 8 steps are very useful in dealing with change for the better. His penguin fable 'Our Iceberg is Melting' (79book) is mandatory reading.

A fitting example of rapid change management is Royal Dutch Shell (86news). In addition to a $4 billion target set in the wake of the Covid-19 crisis, the company is looking to slash up to 40% off the cost of producing oil and gas by the end of 2020, so it can overhaul its business and focus more on renewable energy and power markets.

## Conclusion

Alternative options need to be explored to optimize profits and sustainability. Collective actions by governments, companies, groups and individuals are required to ensure a positive outcome for the world by 2050. Governments and businesses need to compare 'business as usual' models with alternatives to make effective, long term decisions for the future.

# CHAPTER 7: THE FUTURE - WHAT IS LIKELY TO HAPPEN

## Introduction

The Covid-19 pandemic of 2020 has changed the world and it seems fairly certain that purely growth-based economies will not be able to continue as they were, prior to the pandemic. Most businesses will have to align with a new reality and take advantage of sustainable opportunities, those pursuing a 'business as usual' path are unlikely to survive.

One of our greatest challenges is to find ways to feed, employ and accommodate the world's population, anticipated to exceed 9.8 billion by 2050. A greater move towards small and informal businesses appears inevitable.

In this chapter I explore a range of likely scenarios for the future.

## Top countries and companies

According to PricewaterhouseCoopers (PwC), the five top economies in 2030 will be China, India, USA, Indonesia and Brazil and it stands to reason that this is where the foremost companies will be established (1website)(2audio)(3website)(4book).

Tata Industries of India is currently regarded as one of the most ethical companies, although it does not yet rate among the top 25 largest in the world. It has invested heavily in steel production, among other sectors. Since steel production requires a 30 year investment, we can expect to see Tata around in 2030.

The United States has been the global economic powerhouse for decades, although China is rapidly closing the gap. Economists predict that China's economy will surpass the US within the next few years. **Table 7.1** shows the best 10 economies and the predicted number of top companies from these economies.

Table 7.1: Size of Country's Economy relative to Top 25 Companies

| Rank | 1990 | | 2018 | | 2050 PWC | 2050 HSBC | 2050 Author's estimate |
|---|---|---|---|---|---|---|---|
| | Country | Comp. | Country | Comp. | Country | Country | Comp. |
| 1 | US | 10 | US | 9 | China | China | 8 |
| 2 | Japan | 4 | China | 4 | India | US | 4 |
| 3 | Soviet Union | 0 | Japan | 1 | US | India | 6 |
| 4 | W Germany | 3 | Germany | 2 | Indonesia | Japan | 1 |
| 5 | France | 1 | UK | 2 | Brazil | Germany | 1 |
| 6 | UK | 2 | India | 0 | Russia | UK | 1 |
| 7 | Italy | 3 | France | 0 | Mexico | Brazil | 1 |
| 8 | Canada | 0 | Italy | 0 | Japan | Mexico | 1 |
| 9 | Iran | 0 | Brazil | 0 | Germany | France | 1 |
| 10 | Spain | 0 | Canada | 0 | UK | Canada | 1 |

Some top companies in China and India are likely to be in the energy field whereas US companies are more likely to be concentrated in the technology sector. According to PwC, Indonesia, Russia and Mexico will rise to the top 10 and Italy, France and Canada will have lost their positions, although HSBC (61website) has a slightly different view.

I believe that countries that are committed to being carbon neutral by 2050 (34wiki) will be leaders in sustainability. As of 2020 those committed include:

1. Bhutan (carbon negative)
2. Canada
3. Chile
4. Costa Rica
5. Denmark
6. Fiji
7. Finland
8. France
9. Germany
10. Iceland
11. Marshall Islands

12. New Zealand
13. Norway
14. Portugal
15. Sweden
16. UK
17. Uruguay.

The USA, however, has abdicated its leadership role (35book). Since Donald Trump became President in 2017, the USA has:

- Withdrawn from the Paris Accord
- Withdrawn from the UN Human Rights Council
- Initiated a number of trade disputes, the biggest being with China
- Refused to allow the appointment of new judges in the World Trade Organization (WTO) thereby preventing rulings on disputes
- Withdrawn from the Joint Comprehensive Plan of Action (JCPA), also known as the "Iran deal" or the "Iran nuclear deal"
- Withdrawn troops from Northern Syria, leaving the Turks, Russians, Kurds and Syrian Government Forces to fight it out
- Acted tactlessly by using bullying, bellicose and unethical tactics of negotiation (for example in Ukraine)
- Withdrawn funding of the World Health Organization in the middle of the Covid-19 Pandemic.

In May 2019, President Donald Trump issued an executive order banning "foreign adversaries" from US. telecom networks.

> "Removing banned tech from China's Huawei will cost rural Colorado telecoms over $300 million." (36news)

Sense has now prevailed with the 2020 election of Joe Biden as US President. Even as president-elect, he has already stated that he will reverse a number of Trump's executive orders, one of the first being signing the Paris Accord.

After the 2020 Covid-19 virus crisis, the 'apple cart' may well be upturned, giving countries least affected by the virus a boost. The capitalist and communist systems will become less dominant, while community based producer/ consumer 'wellness' economies (37website) will start to blossom, leaving growth-based economies to dwindle.

## Exceeding planetary boundaries

In a moving documentary film, David Attenborough outlines what we have done to destroy planet Earth and what can be done to slow the degradation of Earth's environment (88movie).

A consensus has been agreed on the nine planetary boundaries (13tedtalk) which are:

1. Climate Change - exceeded

2. Rate of Biodiversity loss - exceeded

3. Land system change - exceeded

4. Ocean Acidification

5. Global freshwater use

6. Ozone depletion

7. Atmospheric aerosol loading

8. Chemical pollution

9. Global Nitrogen and Phosphate cycles -exceeded

The first three and last have already been eclipsed. I offer brief comments on just a few of these boundaries and focus on a few likely scenarios here.

### 1. Climate change (exceeded)

The Effect of Global Warming (14website)

*"It is still possible to reduce the rate at which the climate is changing."*

*Prof Ove Hoegh-Guldberg University of Queensland* (81movie)

Climate Action Tracker (31website) is an online tool tracking the progress towards the requirements for 2050. It details by country, pledges and targets, current policy projections and assumptions, and assessment of share of global effort. It also tracks international aviation and marine progress.

The UN Climate Summit 2020 Climate Progress Tracker Tool lists businesses and their commitment to climate action (19website).

**There will be Increasing weather extremes: wild fires, flooding and drought** (104news). BP Statistical Review of World Energy 2019 (38website) has clearly stated that measures to reduce the rate of temperature increase are NOT working.

Green House Gas (GHG) emissions will continue to rise but will hopefully level off after 2030 for the world to become carbon neutral by 2050. **The alternative is unthinkable.**

**Investments in $CO_2$ capture and conversion and in sustainable power are going to increase.**

Carbon footprint awareness and reduction is essential. Some companies are starting to publish their carbon footprint data (39website).

## 2.Rate of biodiversity loss (exceeded)

### Tropical Rainforest Degradation

Tropical rainforest logging will continue. There was a hiatus in Brazil with the previous government, but now the new right wing government is continuing, full steam ahead.

Borneo rain forests continue to be logged with big companies making huge profits while corrupt politicians remain in power. Some indigenous people such as those of the Aru archipelago in Indonesia have stood up to big business (see **Box 7.1**).

> Box 7.1: Saving Aru (42website)
>
> **The epic battle to save the islands that inspired the theory of evolution**
> BY **THE GECKO PROJECT AND MONGABAY** 9 OCTOBER 2019
> *Mongabay Series: Indonesian Fisheries, Indonesian Forests*
>
> *In the mid-1800s, the extraordinary biodiversity of the Aru Islands helped inspire the theory of evolution by natural selection.*
>
> *Several years ago, however, a corrupt politician granted a single company permission to convert most of the islands' rainforests into a vast sugar plantation.*
>
> *The people of Aru fought back. Today, the story of their grassroots campaign resonates across the world as a growing global movement seeks to force governments to act on climate change.*

Degradation of the rain forests is affecting food supply and global warming. To accommodate the need for timber, investments in commercial forestry is likely to grow, although NGOs and indigenous people are going to take back former tropical rain forests and regrow these forests (43tedtalk).

### 3. Land system change (exceeded)

## Beef Production

Cattle farming creates monocultures to feed cattle and **generates up to a quarter of the worlds methane** (15website).

On the up side, a plant based alternative to beef burgers, with the same look, smell and taste, has been developed in a move towards reducing beef production. Even large, predominantly beef based businesses, such as fast food giant Burger King are promoting the alternative. (21website). However the move away from beef farming is going to take a long time.

### 4. Ocean acidification

Ocean degradation will continue with the rise in the earths temperature. We can only hope to slow the damage enough so we can help it to recover. Ocean degradation will reduce a primary source of food and could cause mass starvation.

Proposals include:

- Fishing company alternatives: fish farms
  *Example: Abalone farming*
- Fishing seasons and catch limits
- Coral reef seed bank creation (80website)
- Fisherman awareness programs
  *Example: Turtle escape hatches fitted to fishing nets.*

Investments in fish farming, aquaponics and biodegradable plastic packaging (5website) are going to grow.

## 5. Global freshwater use

Shortages of potable water will become more of an issue. There have already been cases of cities running out of water (99news) or being very close (40website).

Investments in water treatment are undoubtedly going to grow with Abu Dhabi planning to open the largest reverse osmosis water plant in the world in 2022 (41news). New technologies in producing and recycling water are also being developed (98movie).

Damage to the environment when building new water storage dams will need to be carefully assessed with structured, unbiased Environmental Impact Assessments (EIAs).

## 6. Ozone depletion

Ozone depletion is discussed in **Chapter 3 paragraph "Disasters averted or postponed by joint effort".** Scientists expect the Antarctic ozone to recover to the 1980 level around 2070 (49website).

### 7. Atmospheric aerosol loading

In Asia, anthropogenic emissions will continue to increase as urbanization and industrialization proceeds at a breakneck pace. By contrast, aerosols will decline in North America and Europe as factories move to developing countries and Western nations adopt more stringent clean air regulations (25website).

### 8. Chemical pollution

Hopefully the experiences of the past (see **Chapter 3**) will exert pressure on potential polluters. While the focus is on reducing atmospheric emissions and the effect on global warming, chemicals that we use on land will continue to seep into our drinking water, waterways and reach our oceans, rivers and lakes (86website). Chemicals used in gas fracking are already entering underground water resources in the USA (105website).

### 9. Biogeochemical loading: global nitrogen and phosphate cycles

(exceeded)

Modern chemical fertilizers include one or more of the three elements that are most important in plant nutrition: nitrogen, phosphorus, and potassium. Excessive amounts of fertilizer manufactured from fossil fuels are continuing to enter terrestrial and aquatic ecosystems through runoff. Impacts include algae blooms causing the depletion of oxygen in surface waters, pathogens and nitrates in drinking water, and the emission of odors and gases into the air (87website).

## Bribery and corruption

Bribery and corruption will continue to increase as long as there are willing givers and receivers. Democratic processes will attempt to minimize the damage from corruption and whistleblowing will continue to be a major contributor in identifying major cases of corruption. Hopefully large scale corruption will subside.

A company's commitment to establishing and supporting strong anti-bribery and corruption efforts is one of the best ways for an organization to protect itself against reputational, legal, and financial damage (6website).

## Diseases and viruses (9website)

### Announcement

At the Davos World Economic Forum in 2019, the government of the UK announced a five-year plan to tackle the global threat of anti microbial resistance (AMR) (45website). The plan envisages containing and controlling AMR worldwide by 2040.

### Drug abuse

The use of cheap recreational drugs is becoming more frequent with a high possibility of addiction. This has a devastating impact on families and society at large. The youth is being particularly badly affected.

### HIV/AIDS

Up to 20% of the populations of some Southern African countries have HIV/AIDS. Anti-retroviral drugs are expensive, so large parts of poor populations are dying. Awareness programs are starting to reduce the incidences.

### Antibiotics

Antibiotics have had a huge effect in the medical world and have saved many lives, but as the last major class was developed in the 1980's, there has been an increase in the resistance to the medication, primarily due to misuse. Cases of superbug resistance are becoming more and more common, especially in countries with poorer health care. Malawi has a infection epidemic causing sepsis, which nearly killed a fifth of newborns in 2016 (50website).

## Communicable disease control

### Ebola

Ebola is a highly contagious airborne virus. In 2018 Ebola reappeared in the eastern part of the Democratic Republic of Congo where 70% of the world's cobalt, a main ingredient for lithium batteries, is mined. The virus is difficult to control as there are warring factions in the region. The epidemic is still ongoing in 2020.

### Malaria and Dengue

The highest transmission is found south of the Sahara in Africa and in parts of Oceania such as Papua New Guinea. Although some governments are taking preventative measures against these two avoidable diseases, they yet exact a heavy toll on human health.

> *The Malaysian Government takes extensive preventative action against dengue with advertising, fogging and free admittance to government hospitals when diagnosed. Heavy fines are applied on those who have stagnant water harboring mosquito larvae on their properties.*

### Covid-19 pandemic

Covid-19 is having the most devastating effect on the world since World War 2. After the SARS epidemic, various countries took action to establish a rapid response to an outbreak of something similar. To date, Singapore, Taiwan, Hong Kong and New Zealand have been the most successful in preventing the spread of the Covid-19 virus.

**In the post Covid-19 era, the world needs to rethink the direction it is taking, which in the past has been focusing on economic growth. To continue to destroy the natural resources of the world is madness.**

> *During the lockdown, we didn't die without shopping for new clothes every week, and even our shiny car couldn't bring us joy. Instead what did bring happiness was a chat with our loved ones, learning how to*

bake, and maybe just watching the clouds fly by. The consumerism and materialism that fuels carbon emissions and climate change should die along with this virus. (82news)

# The World Wide Web

*"I broadly see three sources of dysfunction affecting today's web:*
*1.Deliberate, malicious intent, such as state-sponsored hacking and attacks, criminal behaviour, and online harassment.*
*2.System design that creates perverse incentives where user value is sacrificed, such as ad-based revenue models that commercially reward clickbait and the viral spread of misinformation.*
*3.Unintended negative consequences of benevolent design, such as the outraged and polarised tone and quality of online discourse"'*
*Tim Berners-Lee Inventor of the World Wide Web* (10website)

## The dark web (7wiki)

The dark web refers to encrypted online content that is not indexed by conventional search engines. The technology routes users' data through a large number of intermediate servers, which protects the users' identity and guarantees anonymity.

Dangers of the dark web include the use of Bitcoin for money laundering, transactions for illegal drugs, and a home for hacking groups, scam sites, illegal porn and terrorism. These threats will continue to grow.

On the positive side, users such as whistleblowers and investigative journalists are protected.

## Big data

Edward Snowden (8movie) made us aware of metadata being mined from Facebook, Google and similar social media and internet sites. As exposed by Snowden, it is being used by intelligence agencies to track down and monitor users of the internet including cell phones and electronic banking (89movie). In spite of these revelations, the tracking is Increasing, keeping up with and including the latest technologies including facial recognition. This is an extremely effective device for security services, but is a serious threat to the privacy of ordinary people.

Personal liberties will come under increasing threat as more and more people use social media. Personal preferences when searching the web have been manipulated by the likes of Google to present alternative results to searches, specifically tailored to the user. This is 'scraped' by or sold to third parties who have used the data to influence the outcomes of referendums and elections (90movie) (91movie).

## Crypto currencies (11website)

Crypto currencies, the most popular being Bitcoin, are based on block chain digital ledgers which use a decentralized, distributed ledger technology that records the origin of a digital asset.

Simple bank transactions using cellphones are flourishing in Sub-Saharan Africa and India while China has Tencent's communications and social platforms for this. Crypto currencies may take this to another level. Paypal has now permitted the use of crypto currencies for transactions (100news).

The People's Bank of China plans to produce a crypto version of the Yuan in the near future (106news).

Facebook has 2.4 billion active users and has proposed their own crypto currency 'Libra' (51video) but US Government approval has not, as yet, been forthcoming.

Concerns about the use of crypto currencies include money laundering and the funding of terrorists and there are even concerns that the global financial system may be disrupted.

## Restoration of biodiversity

Biodiversity is the key to life on Earth and the revival of our damaged planet. Web based tools will increasingly help repair the destruction. Here is an example.

With the use of Google Earth and other tools, The Crowther Lab (74website)(75tedtalk) combines ground data from scientists and practitioners around the world with satellite imagery and environmental information to generate global maps that describe ecosystems. As we enter the UN Decade on Ecosystem Restoration, this Restor platform (97website), an open-data platform of ecological insights, will be used to enable restoration practitioners, land managers and local communities to:
    1. Browse and compare ecological insights
    2. Monitor projects, and
    3. Share information with partners, other restoration organizations, and the public.

## On-line shopping

The web has enabled on-line shopping and, with the Covid-19 induced lock downs around the world, this has grown rapidly and will continue to grow. E-commerce will make up 22% of global retail sales by 2023 (107news).

# Radicalism

Radicalism often leverages legitimate gripes of minority groups and sometimes fill a void when governments are not responsive to the needs of their populations. A classic example is the radicalization of northern Mozambique (see **Chapter 2 Groups: radicals**).

### Containment: The world's policeman has retired (12news).

In 2019 US President Trump decided to withdraw troops from Kurdish

North Syria, allowing the Syrian Government and Russian troops to fill the vacuum. These sort of actions may allow radicalism to flourish again. However, it has been suggested that the withdrawal of US troops from all Muslim countries will deprive radical Muslims of an enemy (60book).

## Continents, countries and cities

### Africa: the dark unknown

Mineral rich Africa needs stability, and corruption to be eliminated, in order to progress.

The Democratic Republic of Congo (DRC) is last on the list of Fragile States and Poorest Countries in the world, although it has many natural resources and abundant hydro-power. With stability and eradication of corruption, DRC would rapidly climb the ladder.

Water is becoming more limited in Sub Saharan Africa. The first city to start the count down to zero water was Cape Town in 2017. Windhoek, the capital of Namibia, has been recycling their potable effluent and blending it with other sources of drinking water since 1968 (52website).

As a result of such water restraints, controls such as of drip irrigation are becoming increasingly important for food production.

Mini electricity grids are mushrooming in large parts of Africa since it's clear that solar photovoltaics will provide the cheapest source of electricity for many of the 600 million people across Africa who don't have access to electricity today (67website).

### Europe

Offshore wind is becoming the dominant player in electricity generation in this region (67website) and may become the foundation for the production of green hydrogen. A Copenhagen consortium is already embarking on an integrated plan to develop a ground-breaking hydrogen and sustainable fuel facility to produce 1,3 GW by 2030, primarily using offshore wind energy (85news).

## Asia

Asian nations such as India, Indonesia, Philippines and Malaysia have growing populations driving economic growth. China's Ambitions are noted in **Box 6.7**.

## Americas

As previously noted, the influence of the US on the world is steadily declining. Democracy is taking hold in many Latin American countries and this, allied to large population growth, should spur the economic wellbeing of these countries.

## C40 cities (54website)

Representing 700+ million citizens and one quarter of the global economy, mayors of the C40 cities are committed to delivering on the most ambitious goals of the Paris Agreement at the local level.

Amsterdam has adopted a holistic City Doughnut approach as a vision and model for shaping the future of the city (see **paragraph 'The age of adaption'**).

# Black swans

Black swans are events that we cannot predict, in other words 'unknown unknowns'.

Black swans include asteroids, tsunamis, volcanoes, disease & virus attacks, terrorist acts etc. What is going to be next? One possibility is that we will reach a tipping point. The Covid-19 virus and the collapse of the oil market are already moving the world towards a tipping point.

# Opportunities that will be exploited

### Electric power generation

According to Bent Erik Bakken chief analyst for DNV GL's Energy Transition Outlook (46website), solar photovoltaics will be the leading source of electricity, providing about 33%, with onshore wind supplying approximately 18%, by 2050. An additional 14% could be provided by hydro-power. Fossil fuels would thus generate 18% and nuclear will provide 4% (76news).

Smart grid technology will be applied extensively to balance supply and distribution in electric power grids.

Coal fired power stations are not going to disappear overnight, making it critical that $CO_2$ capture technology is implemented to reduce emissions from these power stations. Coal fired power generation could also move towards Integrated Gasification Combined Cycle (IGCC) which will drastically reduce emissions (see ***Box 7.2***).

Box 7.2: Integrated Gasification Combined Cycle (IGCC) explained (62website)

*Coal fired power stations typically utilize coal fired boilers to generate steam. The steam drives turbines which connect to alternators that produce electricity. This is referred to as a simple cycle power plant.*

*A combined-cycle power plant uses gas to drive a gas turbine. Hot exhaust gas produces steam in a heat exchanger which in turn drives a steam turbine. Both turbines are connected to alternators which produce up to 50 percent more electricity from the same amount of energy than a traditional simple-cycle plant.*

*IGCC uses a combined cycle format with a gas turbine driven by synthetic gas (syngas) produced from coal gasification, while the exhaust gases are heat exchanged to generate superheated steam to drive a steam turbine. A number of demonstration units, mainly around 250 Mega Watts, are being operated in Europe and the USA.*

*The main incentive for IGCC development has been that units may be able to achieve higher thermal efficiencies than Pulverized Coal Combustion (PCC),* which is common in coal fired power stations today. IGCC will also to match the environmental performance of natural gas-fired plants.

Hydro power generation is being expanded but may be to the detriment of tropical rain forests in some cases (see **Box 5.4**).

Wind and solar power generation is expanding rapidly but the problematic issue of storing energy from these sources is a concern which is being researched intensively. A 200 MW vanadium redox battery (55wiki) has already been developed in China. Excess wind power can be exploited to produce hydrogen. Alternatives to the commonly available polycrystalline solar panels, are being developed, such as Perovskite solar panels which may be twice as efficient and much cheaper (56audio).

Green hydrogen is touted to be the liquid fuel of the future. Using excess wind and solar power, modular hydrogen production units will dot the countryside. Germany has established a national hydrogen strategy which will support green hydrogen production and its adoption across the industrial and transport sectors. Shell is building a large hydrogen plant in north Holland and other energy companies are also investing in hydrogen production.

Israeli and Italian scientists have developed a renewable energy technology that converts solar energy to hydrogen fuel (18website). This would offer a sustainable way to turn water and sunlight into storable energy for fuel cells.

On the nuclear front, Thorium (30website) is being developed as an alternative nuclear fuel. Up and coming nuclear reactor powerhouses China and India both have substantial reserves of Thorium-bearing minerals.

## Process plant

$CO_2$ capture and conversion technology is starting to be implemented in steel mills and other process plants (16news).

Cement manufacturing is a major contributor to global warming. For every ton of cement produced, a ton of $CO_2$ is pumped into the atmosphere. There are some moves to ensure that cement production will be carbon neutral by 2050 (47website). A South African cement company has a novel way of reducing emissions and saving costs as I describe in **Box 7.3**.

Box 7.3: Pretoria Portland Cement's tyre burning (63website)

*In South Africa, Pretoria Portland Cement's de Hoek plant burns waste tyres as an alternative to coal for the purposes of cement production. The project, which enables de Hoek's kiln 6 to burn up to six tyres per minute, reduces the plant's coal use by an estimated 10% while significantly reducing nitrous oxide emissions. It simultaneously decreases landfill requirements for waste tyres, achieving a positive and sustainable environmental impact. The company also blends waste slag from a nearby steelworks to produce 'Slagcem'.*

## Transport power

The use of lithium batteries for powering vehicles is growing exponentially. It is likely that almost 75% of passenger vehicles and 50% of commercial vehicles will be powered by electricity in 2050. Hydrogen fuel cell electric vehicles (FCEVs) will probably become more dominant than current popular battery electric vehicles (BEVs).

In Germany, hydrogen powered fuel cell trains are already being tested as a replacement for diesel powered trains (103news).

For aircraft, the options for alternative propulsion technologies and fuels are limited (77website). Sustainable fuels for jet aircraft (17website) are coming onto the market with lower emissions than conventional aviation fuels (see **Chapter 5 paragraph "Action being taken")**. It's proposed that sustainable aviation fuels (SAF) could achieve 50% of global aviation fuel demand by 2050. This would reduce the life-cycle GHG emissions by 32% compared to conventional fuels. Cryogenic fuels, such as hydrogen, are also being investigated, but safety will obviously be a major issue, the Hindenberg airship disaster being the primary reminder.

According to the International Maritime Organization (IMO), international shipping accounts for 2.2% of global carbon dioxide emissions. IMO aims to halve greenhouse gas emissions from 2008 levels by 2050.

To this end, Denmark's A.P. Moller-Maersk, the world's no. 1 container shipping line, has joined with Norwegian vehicle shipping group Wallenius Wilhelmsen, German car maker BMW, Swedish fashion retailer H&M, British retailer Marks & Spencer and American denim apparel maker Levi Strauss & Co to test an alternative fuel called LEO which is a blend of lignin and ethanol. Lignin, a low-value waste product from ethanol production typically burned for industrial power, can be dissolved in ethanol to create LEO (58news).

Sailing ships might return to the high seas. Danish company, Maersk Tankers, is testing giant rotating cylinders that sit atop its cargo ships, with the hope of cutting fuel use by as much as 10%. The designers of the Vindskip hope they can cut fuel use by 60%, and get the remaining propulsion from natural gas. The ship itself would be the sail; it's tall and relatively thin, like an airplane wing. This would allow the ship to sail more directly into the wind and still generate forward motion. (59news).

Green ammonia is being explored as a future fuel for ships to eliminate harmful GHG emissions (78website). Green ammonia is derived from renewable sources using wind or solar power to produce hydrogen, which is then used to produce ammonia. Blue ammonia, on the other hand, is produced from fossil fuel.

## Recycling

Plastics recycling (20website) is becoming the norm with some plastics even being converted back into liquid fuels. Recycled oil is being converted to bio diesel, and sustainable diesel & aircraft fuel. Hopefully by 2050 nearly 60% of plastics demand could be covered by production based on previously used plastics (33website).

## Biotechnology (22website)

Biotechnology is the use of biological systems found in organisms or the use of the living organisms themselves to make technological advances and adapt those technologies to various fields.

The advent of biotechnology is benefiting fields such as agriculture & animal husbandry, the pharmaceutical industry & medical sciences, and industrial & environmental management.

The results of applying biotechnology in agriculture & animal husbandry include pest resistant crops, and improved plant and animal breeding.

Medical & pharmaceutical products include vaccines and antibiotics. Retroviruses have been discovered and are commonly used to treat HIV/AIDS. DNA (Deoxyribonucleic acid) testing for criminal activities is now common.

The industrial & environmental management uses of biotechnology will become significant in reducing environmental damage. I list a few applications here.

- Fermentation is used to reduce carbon emissions from steel plants (64website) by using anaerobic bacteria to ferment waste emissions to make ethanol for fuel.
- Cleaning up liquid waste and generating power from bio waste uses micro-organisms (65website). Biological treatment is used where micro-organisms in activated sludge break down the pathogens in the waste. Anaerobic digestion is utilized where micro-organisms break down the biodegradable material in the absence of oxygen to produce methane-rich biogas, and bio fertilizer.

## Artificial Intelligence (AI) (23website)

AI is expected to lead the way in the enrichment of human capability. One of the most exciting developments is the enhancement of human techniques in fields such as surgery.

## Nanotechnology

Nanotechnology entails the control of matter on the atomic and molecular scale. Nanotechnology is helping to considerably improve, even revolutionize, many technology and industry sectors: information technology, homeland security, medicine, transportation, energy, food safety and environmental science, among many others.

In the future, nanotechnology could also enable objects to harvest energy from their environment. New nano-materials (66website) and concepts are currently being developed that show potential for producing energy from movement, light, variations in temperature, glucose and other sources with high conversion efficiency.

## Food production

Food production is increasingly coming under pressure. Oceans have been trawled to the point of desertification in places, thereby eliminating any recovery of fish stocks. On land, the scarcity of arable land and limited water availability is going to restrain farming in its present form. In addition, the cultivation of mono cultures does not allow Earth to rejuvenate. There are a few examples where things are slowly changing.

Fish farming in sheltered waters is becoming the norm (24website). The use of dams can be expanded, as shown in the example in ***Box 5.8*** where the dam owner, municipality and trout farm owner have collaborated to create a win-win situation.

Drip irrigation has been successfully exploited in dry climates and the use of hydroponics, aeroponics and aquaponics is becoming more prevalent (5website). An example of a business based on aquaponics is given in ***Box 7.4***.

Box 7.4: Aquaponics Business Example

**'Bowery' the high tech vertical farm** (26website) *(27website)*

The farm of the future uses no soil.

Key elements are:
- Seeds are carefully selected and are not Genetically Modified Organisms (GMO)
- Grown close to home indoors in vertical rows
- Ideal conditions in a controlled indoor growing environment with no pesticides
- Optimal light using LED lights mimicking the full spectrum of the sun.
- Precision farming giving the crops exactly what they need and nothing more: nutrients, water and light.
- Purified water is used to deliver the same nutrients as in highest quality soil. 95% less water than traditional agriculture is used
- Crop cycles are much faster and more frequent than traditional agriculture
- A smarter harvest using sophisticated analytics: crops are harvested at exactly the right time, ensuring flavor is at its prime
- Once it's picked, produce reaches stores and restaurants within a few days, unlike traditional produce which can take weeks.

Tillage of land (plowing) is being found to be highly detrimental to future crops. Regenerative agriculture, involving the principles of no tilling, ensuring crop cover, planting perennials & trees and composting, will become more popular as farmers realize the relative benefits (102movie).

# Other predictions

*"Long term thinking produces better short-term decisions."*

## Gladwell's positive **tipping point** (28website)

Gladwell defines a tipping point as "the moment of critical mass, the threshold, the boiling point". Gladwell states: "Ideas and products and messages and behaviors spread like viruses do". He describes the three "agents of change" in the tipping points of epidemics: connectors, essentially the social equivalent of a computer network hub, maverns who are "information specialists", or "people we rely upon to connect us with new information" and salesmen who are "persuaders", charismatic people with powerful negotiation skills.

This approach could well be used to promote new ideas or replicate proven processes (see also **Chapter 6 paragraph *"Replicating success"***). If novel and exceptional ideas, products or processes can spread like Gladwell's 'viruses', the change to a sustainable carbon neutral world could accelerate dramatically.

So as to implement rapid change in the approach to global warming in the US, the Sunrise Movement, a network of connectors, maverns and salesmen, has studied civil protest, as implemented by the Civil Rights Movement and Gandhi in India, to promote its agenda (109audio). It is hoped that movements, such as this, will spread like wildfire.

## World Economic Forum (WEF) report on jobs

The World Economic Forum (WEF) has produced a report predicting changes in job availability and future skills required (101website). Key findings include the following.

- Automation is increasing requiring reskilling and upskilling, and causing layoffs
- There will be a significant expansion of remote work
- Online learning and training is on the increase
- The public sector needs to provide stronger support for reskilling and upskilling for at-risk or displaced workers.

Jobs on the increase include IT specialists, hands-on technicians, and project & risk managers. Jobs on the decrease relate to routine activities including production line operators, clerks and construction workers.

## Det Norske Veritas' energy modeling

The Energy Transition Outlook (46website) is standards authority Det Norske Veritas' (DNV) view on the energy future through to 2050, modeling 10 regions and the impact on 3 industry sectors. It is a forecast of the most likely path ahead. The model has been re-run for a post Covid-19 environment.

Highlights of the report are summarized as follows:

SHORTER TERM

1. Covid-19 has reduced global energy demand by 8%
2. Energy-related $CO_2$ emissions have peaked and brought forward five years by Covid-19
3. Technology can deliver a Paris-compliant future, if scaled properly
4. Market forces alone will not fix hard-to-abate sectors; stronger policies and regulations are needed.

LONGER TERM

1. Rapid electrification will transform the energy mix by 2050
2. Solar photovoltaic and wind will dominate power generation
3. Natural gas will take over as the largest energy source this decade, and remain so until 2050
4. Despite flat energy demand and a growing renewable share, the energy transition is nowhere near fast enough to deliver on the Paris Agreement. A lot more renewable power, decarbonization, energy-efficiency improvement and carbon capture is needed.

## International Energy Agency (IEA)

The IEA produces a World Energy Outlook (WEO) report annually. The WEO uses a scenario-based approach to highlight the key choices, consequences and contingencies that lie ahead.

> "The world urgently needs to put a laser-like focus on bringing down global emissions. This calls for a grand coalition encompassing governments, investors, companies and everyone else who is committed to tackling climate change."
> Dr Fatih Birol, IEA Executive Director

IEA's World Energy Outlook (WEO) 2020 (67website) presents 3 post Covid-19 scenarios.

## Stated Policies Scenario

In this scenario, global GDP returns to pre-crisis levels in 2021, and global energy demand bounces back by early 2023, but outcomes vary sharply according to fuel type. Renewables meet 90% of the strong growth in global electricity demand over the next two decades, led by continued high levels of solar photovoltaic deployment while global coal use never returns to previous levels.

## Delayed Recovery Scenario

In this scenario global GDP does not recover to pre-crisis levels until 2023, and global energy demand only returns in 2025.

## Sustainable Development Scenario

This scenario sees a near-term surge of investment in clean energy technologies over the next ten years. Along with action to reduce emissions from existing infrastructure, this is enough to make 2019 the definitive peak year for global $CO_2$ emissions.

IEA approximate predictions for the energy mix in 2050 (68website) are:

| | |
|---|---|
| Coal | 12% |
| Oil | 19% |
| Natural Gas | 14% |
| Nuclear | 16% |
| Hydro | 4% |
| Bio and waste | 20% |
| Other incl. geothermal, wind and solar | 15% |

IEA convened its first-ever Clean Energy Transitions Summit (83website) in July 2020 where participants applauded the IEA's post Covid-19 **Sustainable Recovery Plan**, which sets out 30 actionable, ambitious policy recommendations and targeted investments. The Plan, developed in cooperation with the International Monetary Fund, states that 35% of new jobs could be created through energy efficiency measures and another 25% in power systems, particularly in wind, solar and modernizing and strengthening electricity grids.

Here are a few pointers from IEAs Energy Technology Perspective (ETP) 2020 (53website) .

- Transforming the power sector alone will only get us one-third of the way to net-zero emissions
- The emissions from power, iron & steel, and cement factories will be with us for many years
- The emission issue of the world's existing infrastructure needs to be addressed.

*"There are still 'huge challenges' to meeting Paris Agreement emissions targets that will require greater levels of technological development and government support to overcome,.."*

*"A key issue is for the governments and investors to make the right technology policies and make the right technology investments. This is extremely critical," IEA Chef Birol at the launch of the IEA's Energy Technology Perspectives (ETP) report  (92website)*

## PricewaterhouseCoopers (PwC): The World in 2050 (1website)

PwC has attempted to predict how the global economic order will change by 2050. The company's analysis includes a number of key challenges for policy-makers:

- Avoid a slide back into protectionism, which history suggests would be bad for global growth in the long run
- Ensure that the potential benefits of globalization are shared more equally across society

- Develop new green technologies to ensure that long-term global growth is environmentally sustainable.

## World Bank Group: The Road to 2050 (29website)

The Road to 2050 was published in 2006 and emphasizes five issues that will be critical in achieving the vision of **A Wealthier More Equitable World by 2050.**

1. Sustaining natural wealth
2. Improving governance
3. Achieving social development
4. Boosting agricultural productivity and competitiveness
5. Managing climate risks.

The report states that poor governance is a major constraint and rectifying this will require both institutional reforms and the mobilization of civil society. In addition human capital, energy, infrastructure, agricultural productivity, aid and trade will be essential underpinnings for achieving a wealthier, more equitable world by 2050.

There are some predictions that the situation will improve.

- Basic human needs for shelter, food, and clothing could be more than met. People would be healthier and more skilled.
- Life expectancy in 2050 will be 72 years in low- and middle-income countries (up from 64 today)
- Mortality rates for under 5's will be at 17 per 1,000 live births (down from 85 per 1,000 today).
- Adult illiteracy rates could be less than 5%, a fifth of today's 25%.

These predictions are supported by statistics showing that things have improved (69book) and thus should continue this trend in the long term.

## HSBC: The World in 2050 (32website)

HSBC published The World in 2050 in November 2019. Their key points are:

1. Nineteen of the 30 largest economies will be emerging economies

2. The emerging economies will collectively be bigger than the developed economies

3. Global growth will accelerate thanks to the contribution of the emerging economies.

According to HSBC, Asia will continue demonstrating extremely strong growth rates and those with large populations will overtake western powerhouses. Latin America will feature more heavily while the smaller European countries may struggle to maintain their influence in global policy forums. China and India will be the largest and third-largest economies in the world, respectively. Emerging economies will make substantial progress, most notably, Mexico, Turkey, Indonesia, Egypt, Malaysia, Thailand, Colombia and Venezuela.

HSBCs predictions were published before the Covid-19 pandemic emerged, although they state that natural disasters could send economies seriously off course as their development seeks to replace what was lost rather than make any further forward progress. This may well happen while recovering from the Covid-19 pandemic, but it's too early, as yet, to foresee the long term effects of the pandemic.

## The Millennium Project (MP)

The Millennium Project is a global participatory think tank, established in 1996 under the American Council for the United Nations University (70website), which has grown to 63 Nodes around the world (an MP Node is a group of institutions and individuals that connect local and global perspectives).

The purpose of the MP is to assist in organizing futures research by continuously updating and improving humanity's thinking about the future. The MP manages processes that collect and assess data from its several hundred participants to produce the annual "State of the Future" series, and special studies.

A major outcome has been the presentation of 15 global challenges to provoke thought on how to address them. I've listed them in **Box 7.5**.

### Box 7.5: Millennium Project's 15 global challenges

1. How can sustainable development be achieved for all while addressing global climate change?
2. How can everyone have sufficient clean water without conflict?
3. How can population growth and resources be brought into balance?
4. How can genuine democracy emerge from authoritarian regimes?
5. How can decisionmaking be enhanced by integrating improved global foresight during unprecedented accelerating change?
6. How can the global convergence of information and communications technologies work for everyone?
7. How can ethical market economies be encouraged to help reduce the gap between rich and poor?
8. How can the threat of new and reemerging diseases and immune micro-organisms be reduced?
9. How can education make humanity more intelligent, knowledgeable, and wise enough to address its global challenges?
10. How can shared values and new security strategies reduce ethnic conflicts, terrorism, and the use of weapons of mass destruction?
11. How can the changing status of women help improve the human condition?
12. How can transnational organized crime networks be stopped from becoming more powerful and sophisticated global enterprises?
13. How can growing energy demands be met safely and efficiently?
14. How can scientific and technological breakthroughs be accelerated to improve the human condition?
15. How can ethical considerations become more routinely incorporated into global decisions?

The above subjects are presented in YouTube *(71video)*.

# The age of adaption

## Moving from capitalism and communism to wellbeingism (73video)

With mass unemployment after the Covid-19 pandemic, countries will need to think of adapting established political models, and companies will have to become more inclusive. We will have to embrace creative adaptation to find ways to reboot our societies and reinvent the way in which our economies work. We need to focus on wellbeingism and possibly measuring this with a Well Being index for each country (79book). There is currently a wellbeing government alliance with member countries which include Scotland, Iceland, New Zealand and Wales (48website).

## Building back better after the Covid-19 pandemic

> "Governments have a once-in-a-lifetime opportunity to reboot their economies and bring a wave of new employment opportunities while accelerating the shift to a more resilient and cleaner energy future."
>
> EIA chief Birol (84news)

**A return to 'business as usual' and environmentally destructive investment activities must be avoided at all cost,** for the economic recovery from the Covid-19 crisis to be durable and resilient.

Building back better (72website) requires that recovery policies need to trigger investment and behavioural changes that will reduce the likelihood of future shocks and increase society's resilience to them when they do occur. This approach requires a focus on wellbeing (93website) and inclusiveness, alignment with long-term emission reduction goals, resilience to climate impacts, slowing of biodiversity loss and increasing circularity of supply chains.

The use of Gross Domestic Product (GDP) should no longer be used as the prime indicator of holistic improvement for a country.

> *"On 12th August 2020, the Office for National Statistics announced that the UK's GDP had fallen 20.4% in the second quarter, putting the UK into its worst recession since records began".* (94news)

A key tool in building back better is the application of doughnut economics as proposed by Kate Raworth (95book)(96website). The doughnut is based on the framework of planetary boundaries (see **Chapter 5 Introduction**) and the demands of social justice. It brings social and environmental concerns together in one single image and approach, acting as a convening space for debating alternative ways forward. In 2019 the approach was piloted in three cities: Philadelphia, Portland and Amsterdam.

> *"A healthy economy should be designed to thrive, **not grow**."* (108tedtalk)

## Conclusion

Conflicting forces are at play. Will radicalism and corruption win, or will ethical and civil behavior dominate? The pendulum of democracy is swinging between extremes: leftist chaos and the neo-facisism of the right. Can countries recover from extreme politics?

In this book, I have outlined the past, present and what has, or what has not, been learnt. Improved policies, controls and actions are necessary to get to a carbon neutral world by 2050. Key factors and measurement of aspects of these factors have been proposed. Various scenarios are outlined with emphasis on achieving the requirements of the Paris Accord. This can only happen once the world turns to a circular economy, where what is taken from the environment is replenished, with the primary objective being to become carbon neutral.

The Covid-19 pandemic of 2020 has caused a major drop in GHG emissions with reduced road, sea and air activity. This has been an opportunity for us to pause and consider the direction we are taking in destroying our planet.

The concept of a wellbeing economy needs to be explored and rapidly developed as the growth economy of the past will fail. Networks of local producers/consumers need to be developed to prevent starvation. These networks were demonstrated during the Covid-19 'lockdown' where international shipping was drastically reduced.

Businesses need to reset their objectives to ensure we all survive beyond the 100 years predicted by 'The Limits to Growth'.

We have 30 years for the world to achieve a zero carbon environment. This is just one of the objectives that we have to achieve to survive.

For humanity to endure, 'business as usual' must DIE.

# References

## Introduction

1. Meadows DH, Meadows DL, Danders J, Behrens WW The Limits to Growth Universe Books 1972 https://www.publicspaceinfo.nl/media/uploads/files/CLUBVANROM_1972_0001.pdf

## Chapter 1: Factors for a Sustainable Successful Company

1. Semler Tedtalk https://www.ted.com/talks/ricardo_semler_how_to_run_a_company_with_almost_no_rules
2. Glassdoor best places to work https://www.inc.com/peter-economy/glassdoor-just-announced-100-best-places-to-work-for-2019-is-your-company-on-list.html
3. Branson R The Virgin Way page 163 Virgin 2014 https://www.amazon.com/Virgin-Way-Its-Worth-Doing/dp/1591847982
4. Semler R Maverick Arrow Books 1993 https://www.amazon.com/Maverick-Success-Behind-Unusual-Workplace/dp/0446670553
5. Chobani yogurt https://www.ft.com/content/eda437f8-76db-11e9-be7d-6d846537acab
6. Hamdi Ulukaya The anti-CEO playbook Tedtalk April 2019 https://www.ted.com/talks/hamdi_ulukaya_the_anti_ceo_playbook?language=en
7. Sseairtricity http://www.sseairtricity.com/ie/home/ and https://www.youtube.com/channel/UCG2XXPipczgYtcmBXkO1Xtw
8. Mainstream https://www.mainstreamrp.com/what-we-do/#corporate-solutions
9. Microsoft Social Responsibility https://www.geekwire.com/2019/microsoft-approaching-seattle-regions-affordable-housing-crisis/
10. Ikea Foundation https://ikeafoundation.org/
11. Volkswagen Annual Report 2018 https://annualreport2018.volkswagenag.com/group-management-

report/shares-and-bonds/shareholder-structure.html
12. New York Times, Sreeharsha V. Brazilian senator and banker are arrested as Petrobras scandal widens. November 25, 2015 https://www.nytimes.com/2015/11/26/world/americas/brazil-corruption-petrobas.html
13. Tran M. Shell fined over reserve scandal. The Guardian; July 29, 2004 https://www.theguardian.com/business/2004/jul/29/oilandpetrol.news
14. Shell and the SEC CNN/Money. http://money.cnn.com/2004/08/24/news/international/royaldutchshell_sec
15. Du Pont class action documentary film 'The Devil We Know- The Chemistry Of A Cover Up' 2018 https://www.youtube.com/watch?v=NJFbsWX4MJM
16. Daily Sabah French groups sue Samsung over child labour abuse Jan 2018 https://www.dailysabah.com/technology/2018/01/11/french-groups-sue-samsung-over-child-labor-abuse
17. New York Times The Daily podcasts 'Boeing' March 19, April 23, July 30 2019 https://www.nytimes.com/2019/07/30/podcasts/the-daily/boeing-737-max.html
18. Glencore Wikipedia https://en.wikipedia.org/wiki/Glencore
19. Hydrocarbon Processing 'Glencore's head of oil, Alex Beard, retires amid U.S. probes' 6/3/2019 https://www.hydrocarbonprocessing.com/news/2019/06/glencores-head-of-oil-alex-beard-retires-amid-us-probes
20. Hornbillunleashed Fact Based Communication (FBC) https://hornbillunleashed.wordpress.com/2011/10/28/24847/
21. New York Times Bell Pottinger https://www.nytimes.com/2017/09/12/business/bell-pottinger-administration.html
22. Marketing Interactive Bell Pottinger Asia https://www.marketing-interactive.com/bell-pottinger-asia-name-change-good-pr-move-for-a-distressed-pr-firm/
23. SLC Cambridge Analytica documentary film 'The Great Hack' Netflix 2019 https://www.netflix.com/za/title/80117542
24. Big Pharma Netflix documentary film 'The Bleeding Edge' Dick K 2018 https://www.netflix.com/za/title/80170862

25. BBC Johnson & Johnson https://www.bbc.com/news/business-44816805
26. Attorney General Washington State Johnson & Johnson https://www.atg.wa.gov/news/news-releases/johnson-johnson-will-pay-99-million-failing-disclose-risk-its-surgical-mesh
27. The Guardian Contraceptive implant https://www.theguardian.com/society/2018/nov/25/contraceptive-implant-essure-surgically-removed-from-thousands-women
28. Alaska Public artificial joints https://www.alaskapublic.org/2016/06/03/cobalt-poisoning-from-hip-replacement-surgery/
29. The Guardian US opioids pandemic https://www.theguardian.com/us-news/2020/jul/09/coronavirus-pandemic-us-opioids-crisis
30. Opiods crisis Freakonomics 15 Jan 2020 'Opiod Tragedy' https://freakonomics.com/podcast/opioids-part-1/
31. Tobacco industry settlement Wikipedia https://en.wikipedia.org/wiki/Tobacco_Master_Settlement_Agreement
32. Samsung alleged child labour abuses https://www.business-humanrights.org/en/france-samsung-electronics-indicted-for-misleading-advertising-re-alleged-labour-abuses-child-labour-in-china-s-korea-vietnam
33. Organization for Economic Cooperation and Development OECD https://www.oecd.org/
34. Bribery 1997 OECD Convention on Combating Bribery.
35. Extractive Industries Transparency Initiative EITI https://eiti.org/
36. Reuters Ichan Occidental dispute https://www.reuters.com/article/us-occidental-icahn/icahn-steps-up-fight-with-occidental-over-deal-wants-board-seats-idUSKCN1TR36Y?dlbk
37. Garratt B. 'The Fish Rots from the Head: : The Crisis in Our Boardrooms: Developing the Crucial Skills of the Competent Director' pg 211 Profile Books 2010 https://www.amazon.com/Fish-Rots-Head-Boardrooms-Developing/dp/1846683297
38. Garratt B. 'Stop the Rot: Reframing Governance for Directors and Politicians' 2017 https://www.amazon.com/Stop-Rot-Reframing-Governance-Politicians/dp/1783537663
39. ASEAN https://www.britannica.com/event/Asian-financial-crisis
40. Goedgedacht www.goedgedacht.org

41. Crane and Matten 'Business Ethics: Managing Corporate Citizenship and Sustainability in the Age of Globalization' Oxford University Press 2016 https://www.amazon.com/Business-Ethics-Citizenship-Sustainability-Globalization/dp/0199697310
42. Business Round Table(BRT) policy revision https://www.theatlantic.com/ideas/archive/2019/08/milton-friedman-shareholder-wrong/596545/
43. New York Times The Daily podcast Michael Barbaro 21/08/19 'What American CEOs are worried about' https://www.nytimes.com/2019/08/21/podcasts/the-daily/business-roundtable-corporate-responsibility.html
44. Business Round Table(BRT) https://www.businessroundtable.org/business-roundtable-redefines-the-purpose-of-a-corporation-to-promote-an-economy-that-serves-all-americans
45. UN sustainable development https://sustainabledevelopment.un.org/sdgs
46. World Business Council for Sustainable Development (WBCSD) https://www.wbcsd.org/
47. UN global compact https://www.globalcompact.ca/about/ungc-10-principles/
48. Principles of responsible investment UN PRI https://www.unpri.org/pri/what-are-the-principles-for-responsible-investment
49. Harvard Business Review The Investor Revolution Robert G. Eccles, Svetlana Klimenko from the May–June 2019 Issue Sustainable Development Projects https://hbr.org/2019/05/the-investor-revolution
50. Dorling Kindesley 'Whats where in the world' page 33 https://www.dk.com/uk/book/9781409379249-whats-where-in-the-world/
51. International Sustainability and Carbon Certification (ISCC) https://www.iscc-system.org/about/objectives/
52. Borealis https://www.hydrocarbonprocessing.com/news/2020/03/borealis-producing-certified-renewable-polypropylene-from-neste-s-renewable-propane-at-own-facilities-in-belgium?id=3872815
53. Forestry Stewardship Certificate (FSC)

https://www.scsglobalservices.com/fsc-mark-of-responsible-forestry
54. Programme for the Endorsement of Forest Certification (PEFC) www.pefc.org
55. Sustainable forest management - Wikipedia https://en.wikipedia.org/wiki/Sustainable_forest_management
56. Carbon neutral certificate https://www.carbontrust.com/what-we-do/assurance-and-certification/carbon-neutral-certification
57. 'Gasland' IMDB documentary film 2010 https://www.imdb.com/title/tt1558250/
58. Bruno Manser Fund http://bmf.ch/en
59. News / 16 November 2016 Client Earth First shipment ceremony FLEGT Licence Indonesia, https://www.clientearth.org/first-shipment-licensed-legal-timber-indonesia/
60. Client Earth https://www.clientearth.org
61. Center for International Forestry Research (CIFOR) https://www.cifor.org/
62. South Korea political scandal https://www.bbc.com/news/world-asia-37971085
63. Eskom Kusile project Biznews https://www.biznews.com/energy/2019/07/29/crisis-kusile-eskom-chris-yelland
64. Texaco Ecuador rainforest disaster Documentary film 'Crude' IMDB 2009 https://www.imdb.com/title/tt1326204/
65. BP and PT Freeport https://newint.org/features/2017/05/01/sacrifice-zone-west-papuan-independence-struggle
66. Exploring Corporate Strategy Johnson, Scholes & Whittington pg144 https://www.amazon.com/Exploring-Corporate-Strategy-Gerry-Johnson/dp/0273711911
67. Lockheed Scandal https://www.washingtonpost.com/archive/business/1977/05/27/lockheed-paid-38-million-in-bribes-abroad/800c355c-ddc2-4145-b430-0ae24afd6648/
68. Public Protector wikipedia https://en.wikipedia.org/wiki/Public_Protector
69. Organized Crime and Corruption Reporting Project (OCCRP) South African Public Protector https://www.occrp.org/en/investigations/south-africas-public-protector-flagged-by-hsbc-in-connection-to-guptas

70. Corruption Perception Index (CPI) www.transparency.org
71. OECD Anti Bribery Convention wikipedia https://en.wikipedia.org/wiki/OECD_Anti-Bribery_Convention
72. G20 anti corruption working group https://www.oecd.org/g20/topics/anti-corruption/
73. Transparency International anti bribery principles https://www.transparency.org/whatwedo/tools/business_principles_for_countering_bribery/1
74. ISO 20000 social responsibility https://www.iso.org/standard/42546.html
75. ISO 37001 anti bribery https://www.iso.org/standard/65034.html
76. Segal N Breaking the Mould SUN press 2007 https://scholar.sun.ac.za/handle/10019.1/101782
77. The Sun asteroid collisions https://www.thesun.co.uk/wp-content/uploads/2019/06/pmasteroidtablev5.jpg
78. Bcorps certification https://bcorporation.net/about-b-corps
79. VW ethics 'Hard NOx - the Volkswagen emissions scandal' Dirty Money season 1 no 1 2018 Netflix https://www.netflix.com/za/title/80118100
80. Nigeria SEC https://www.petroleum-economist.com/articles/politics-economics/africa/2019/oando-under-fire-from-nigerias-sec
81. ISO 22301:2012, Business continuity management https://www.iso.org/standard/50038.html
82. Investopedia SEC https://www.investopedia.com/terms/s/sec.asp
83. Rewcastle-Brown Sarawak Report http://www.sarawakreport.org/about/
84. IKEA Foundation tents https://www.dezeen.com/2017/04/29/united-nations-admits-10000-ikea-better-shelter-refugees-mothballed-fire-fears/

## Chapter 2: Influencers

1. Harvard Business Review investor revolution https://hbr.org/2019/05/the-investor-revolution
2. Petroleum Economist Investors and big oil https://www.petroleum-economist.com/articles/low-carbon-energy/renewables/2019/the-climate-threat-will-investors-exit-big-oil
3. Shell shareholder agreement https://www.shell.com/media/news-and-media-releases/2018/joint-statement-between-institutional-investors-on-behalf-of-climate-action-and-shell.html
4. Blackrock ...https://www.blackrock.com/corporate/investor-relations/larry-fink-ceo-letter
5. Vanguard https://www.eco-business.com/opinion/blackrock-is-getting-serious-about-climate-change-is-this-a-turning-point-for-investors/
6. State Street https://www.statestreet.com/values/environmental-sustainability.html
7. Freakonomics radio 13 August 2020 Episode 429 Is Economic Growth the Wrong Goal? https://podcasts.google.com/feed/aHR0cHM6Ly9hcHAuYWxwaGF2b2ljZS5pby9wb2RjYXN0LzQ0LWZyZWFrb25vbWljcy9mZWVk/episode/M2M0MTAyZDktNzNiMS00MmFmLTgzM2EtYWMxNTAxNzQ5ZjA4?hl=en-ZA&ved=2ahUKEwjt55Xe4dnrAhXxoFwKHTLfBwsQieUEegQINxAU&ep=6
8. Freedom House https://freedomhouse.org/
9. Transparency International https://www.transparency.org/cpi2018
10. Ibrahim Prize https://en.wikipedia.org/wiki/Ibrahim_Prize
11. Ibrahim Index https://www.africanexponent.com/post/9335-mauritius-tops-the-ibrahim-index-of-african-governance-2018-ranking
12. PWC https://www.pwc.com/gx/en/world-2050/assets/pwc-the-world-in-2050-full-report-feb-2017.pdf
13. Yes magazine Costa Rica Happiness Index https://www.yesmagazine.org/issue/climate-action/2019/01/31/why-costa-rica-tops-the-happiness-index/
14. Germany & China solar and wind Dorling Kindesley 'Whats where in the world' page 33 https://www.dk.com/uk/book/9781409379249-whats-where-in-the-world/
15. Resource curse wikipedia

https://en.wikipedia.org/wiki/Resource_curse#:~:text=The%20resource%20curse%2C%20also%20known,countries%20with%20fewer%20natural%20resources.

16. Top Corruption Scandals https://www.transparency.org/news/feature/25_corruption_scandals
17. Oil related crimes https://oilprice.com/Latest-Energy-News/World-News/Nigeria-Lost-28B-To-Oil-Related-Crimes-Last-Year.html
18. ENI and Shell court action https://www.petroleum-economist.com/articles/politics-economics/africa/2019/eni-and-shell-face-new-nigeria-action
19. Angola oil revenues https://www.hrw.org/report/2004/01/12/some-transparency-no-accountability/use-oil-revenue-angola-and-its-impact-human
20. IMF Country Report No. 18/157 Angola https://www.imf.org/en/Publications/CR/Issues/2018/06/11/Angola-Selected-Issues-45958
21. Lake Tekana Fishing Project and other failed aid funded projects http://www.nbcnews.com/id/22380448/ns/world_news-africa/t/examples-failed-aid-funded-projects-africa/#.XPcwMo8RVPY
22. World Bank http://www.worldbank.org/en/news/feature/2013/04/17/ending_extreme_poverty_and_promoting_shared_prosperity
23. Water project failure in Tanzania https://www.pri.org/stories/2014-11-24/world-banks-water-failure-tanzania
24. Kenya Off Grid Solar http://projects.worldbank.org/P160009?lang=en
25. Regional Off Grid Electrification Project http://projects.worldbank.org/P160708?lang=en
26. World Bank https://www.worldbank.org/en/about/partners/maximizing-finance-for-development#6
27. African Development Bank https://www.afdb.org/en/about/mission-strategy
28. Asian Development Bank https://www.adb.org/
29. Goldman Sachs Bank https://www.straitstimes.com/asia/se-asia/goldmans-1mdb-case-in-malaysia-to-be-moved-to-higher-court
30. BSI Bank https://theconversation.com/how-a-swiss-bank-was-toppled-by-a-financial-scandal-in-malaysia-and-what-can-be-learned-from-it-100130

31. BSI EFG Banks https://www.finews.com/news/english-news/26969-bsi-banca-svizzera-italiana-efg-private-bank-wealth-management-private-banking-asia-1mdb
32. Ambank http://www.sarawakreport.org/2019/01/deafening-silence-out-of-australia-over-1mdbs-connection-to-top-bank-anz/
33. J P Morgan Bank https://www.reuters.com/article/us-jpmorgan-swiss-1mdb/swiss-find-serious-shortcomings-at-jpmorgan-in-1mdb-case-idUSKBN1EF13T
34. Standard Chartered, Coutts Banks https://www.reuters.com/article/us-malaysia-scandal-singapore/singapore-slaps-penalties-on-stanchart-coutts-in-1mdb-related-probe-idUSKBN13R06F
35. Coutts Bank https://www.straitstimes.com/business/swiss-fine-coutts-over-breaches-linked-to-1mdb
36. Deutche Bank https://www.nst.com.my/news/nation/2019/07/503442/us-probes-deutsche-banks-dealings-malaysias-1mdb-wsj
37. DBS UBS Falcon Bank https://www.straitstimes.com/business/swiss-banks-chasing-asia-cash-stung-by-dirty-money-crackdown
38. Falcon Bank https://fortune.com/2016/10/11/1mdb-singapore-dbs-ubs-falcon-bank/
39. Rothchild Bank https://www.bloomberg.com/news/articles/2018-07-20/rothschild-bank-broke-money-laundering-rules-in-1mdb-case
40. Tories ensure London remains the money laundering capital of the world Sky News https://www.youtube.com/watch?v=EtvXl8ZuPU4
41. New York Times Art and money laundering https://www.nytimes.com/2017/02/19/arts/design/has-the-art-market-become-an-unwitting-partner-in-crime.html
42. WEF Closing the Skills Gap Initiative http://www3.weforum.org/docs/WEF_System_Initiative_Future_Education_Gender_Work_Closing_Skills_Gap_pagers....pdf .
43. Cambodia cafe management https://www.cam.ac.uk/footprintcafe
44. Cambodia cookery school https://havencambodia.com/about-us/our-story/
45. UK NQF https://en.wikipedia.org/wiki/National_qualifications_frameworks_in_the_United_Kingdom

46. NZ NQF https://www.nzqa.govt.nz/studying-in-new-zealand/understand-nz-quals/nzqf/
47. SA NQF https://www.saqa.org.za/
48. National Qualifications Frameworks https://en.wikipedia.org/wiki/National_qualifications_framework
49. Forbes General Motors Bankruptcy https://www.forbes.com/sites/danbigman/2013/10/30/how-general-motors-was-really-saved-the-untold-true-story-of-the-most-important-bankruptcy-in-u-s-history/?sh=1bddaee97eea
50. Sarawak Report http://www.sarawakreport.org/
51. Bellingcat Podcast: MH17, Episode 4 Guide: Manhunt https://www.bellingcat.com/resources/podcasts/2019/08/07/bellingcat-podcast-mh17-episode-4-guide-manhunt/
52. Organized Crime and Corruption Reporting Project https://www.occrp.org/en/about-us
53. Bureau of Investigative Journalism https://www.thebureauinvestigates.com/
54. The Guardian Superbug hot spots https://www.theguardian.com/environment/2019/sep/19/superbug-hotspots-emerging-in-farms-across-globe-study?CMP=Share_iOSApp_Other
55. The Centre of Investigative Journalism (TCIJ) https://tcij.org/
56. Reporters Sans Frontières https://rsf.org/en
57. Human Rights Watch https://www.hrw.org/world-report/2019
58. Actionaid https://actionaid.org/
59. Eartheasy https://learn.eartheasy.com/articles/where-to-donate-10-high-impact-environmental-charities-with-integrity/
60. Masarang https://masarang.nl/en/
61. Environmental Investigation Agency (EIA) https://eia-international.org/about-us/what-we-do/
62. End Plastic Waste https://endplasticwaste.org/
63. Plastic Tides https://www.worldbank.org/en/news/feature/2019/06/04/meet-the-innovator-battling-plastic-waste-in-the-philippines-julian-rodriguez
64. Tedtalk Sea Sheperds https://www.youtube.com/watch?v=Y0XOx_UVRPo

65. The Daily Podcast 4 June 2019: How a secret US cyberweapon backfired. https://www.nytimes.com/2019/06/04/podcasts/the-daily/nsa-hacking-tool-baltimore.html
66. Scahill J Blackwater: The Rise of the World's Most Powerful Mercenary Army Nation Books 2007 https://www.amazon.com/Blackwater-Rise-Worlds-Powerful-Mercenary-ebook/dp/B0097CYTYA
67. Wagner private army https://www.youtube.com/watch?v=F5VvLF0WVeY
68. Freeport and the Indonesian security forces https://reliefweb.int/report/indonesia/paying-protection-freeport-mine-and-indonesian-security-forces
69. Freemasons https://en.wikipedia.org/wiki/Freemasonry
70. Human Rights Watch https://www.hrw.org/news/2019/09/22/indonesia-indigenous-peoples-losing-their-forests
71. Bruno Manser Fund  https://bmf.ch/en
72. International Work Group for Indigenous Affairs https://www.iwgia.org/en/
73. Nissan UK productivity https://www.just-auto.com/news/nissans-sunderland-plant-is-europes-most-productive-reports-just-autocom_id75827.aspx
74. GM strike https://www.wsj.com/articles/gm-strike-heads-into-a-second-week-11569201210
75. Guardian's Audio Long Reads 11 Dec 2019 'How the Right's radical think tanks reshaped the Conservative Party' https://www.theguardian.com/politics/audio/2019/dec/11/how-the-rights-radical-thinktanks-reshaped-the-conservative-party-podcast
76. World Economic Forum https://en.wikipedia.org/wiki/World_Economic_Forum
77. G7 countries https://en.wikipedia.org/wiki/Group_of_Seven
78. G20 countries http://worldpopulationreview.com/countries/g20-countries/
79. Bill and Melissa Gates Foundation (BMGF) https://www.gatesfoundation.org/
80. The Howard G Buffett Foundation https://www.thehowardgbuffettfoundation.org/about/
81. Thurow R and Kilman S. Enough: Why the World's Poorest Starve in an Age of Plenty https://www.amazon.com/Enough-Worlds-Poorest-Starve-

Plenty/dp/158648818X
82. The Open Society https://www.opensocietyfoundations.org/
83. Grimeen Foundation https://grameenfoundation.org/
84. Yunus Social Business https://www.yunussb.com/
85. Berners-Lee https://www.w3.org/People/Berners-Lee/
86. Wikipedia https://en.wikipedia.org/wiki/Main_Page
87. Saro-Wira and Shell https://www.telegraph.co.uk/news/worldnews/africaandindianocean/niger/5413171/Shell-execs-accused-of-collaboration-over-hanging-of-Nigerian-activist-Ken-Saro-Wiwa.html
88. The Guardian Thunberg https://www.theguardian.com/world/2019/mar/11/greta-thunberg-schoolgirl-climate-change-warrior-some-people-can-let-things-go-i-cant
89. Rewcastle-Brown Sarawak Report http://www.sarawakreport.org/about/
90. Documentary film 'Bellingcat: Truth in a Post Truth World' IMDB 2018 https://www.imdb.com/title/tt7844518/
91. Vance A 'Elon Musk: Tesla, SpaceX, and the Quest for a Fantastic Future' Publisher 2017 https://www.amazon.com/Elon-Musk-SpaceX-Fantastic-Future/dp/006230125X
92. Carrerou J. 'Bad Blood: Secrets and Lies in a Silicon Valley Startup' Alfred A Knopf 2018 https://www.amazon.com/Bad-Blood-Secrets-Silicon-Startup/dp/152473165X
93. 'The Inventor-Out for Blood in Silicon Valley' documentary film IMDB 2019 https://www.imdb.com/title/tt8488126/
94. Guptas https://en.wikipedia.org/wiki/Gupta_family
95. Sarawak Report http://www.sarawakreport.org/
96. Najib Razak https://en.wikipedia.org/wiki/Najib_Razak
97. Zuma https://en.wikipedia.org/wiki/Jacob_Zuma
98. Dilma Rousseff The Guardian https://www.theguardian.com/news/2016/aug/31/dilma-rousseff-impeachment-brazil-what-you-need-to-know
99. Murdoch https://en.wikipedia.org/wiki/Rupert_Murdoch
100. Documentary film 'Rogue Trader' IMDB 1999 https://www.imdb.com/title/tt0131566/
101. Leeson and Barings https://www.investopedia.com/ask/answers/08/nick-

leeson-barings-bank.as
102. Skilling and Enron https://www.britannica.com/event/Enron-scandal
103. Jo Low 1MDB https://www.businesstimes.com.sg/government-economy/malaysia-sought-hong-kongs-help-in-hunt-for-1mdb-fugitives
104. Bernie Madoff https://en.wikipedia.org/wiki/Madoff_investment_scandal
105. Panama Papers https://en.wikipedia.org/wiki/Panama_Papers
106. Sarawak Report http://www.sarawakreport.org/2016/09/held-hostage-najib-pressures-thailand-to-keep-star-witness-xavier-justo-jailed-exclusive/
107. 'Snowden' IMDB documentary Oliver Stone 2016 https://www.imdb.com/title/tt3774114/
108. BNP Petroleum Economist https://www.petroleum-economist.com/articles/low-carbon-energy/energy-transition/2020/asset-managers-deliver-brutal-verdict-on-oil
109. Investors and NGOs take tougher line on climate change Petroleum Economist https://www.petroleum-economist.com/articles/low-carbon-energy/energy-transition/2020/investors-and-ngos-take-tougher-line-on-climate-change
110. IMDB documentary series Dirty Money Cartel Bank episode 2018 https://www.imdb.com/title/tt7909188/
111. Aldrich Ames https://en.wikipedia.org/wiki/Aldrich_Ames
112. Robert Hanson https://en.wikipedia.org/wiki/Robert_Hanssen
113. Huawei ban https://www.cnet.com/news/huawei-ban-full-timeline-mate-30-pro-security-threat-china/
114. Shell class action https://www.foei.org/press_releases/climate-legal-summons-submitted-shell
115. G20 protests Japan coal http://www.nocoaljapan.org/g20-protests-japan-end-coal/
116. Duke of Edinburgh Award www.intaward.org
117. International Civil Aviation Organization (ICAO) https://www.icao.int/about-icao/Pages/default.aspx
118. 'Fear City - New York verses the Mafia' IMDB documentary 2020 https://www.imdb.com/title/tt12588372/
119. BBC Goldman Sachs 1MDB settlement https://www.bbc.com/news/business-53529075

120. International Maritime Organization (IMO) http://www.imo.org/en/About/Pages/Default.aspx
121. Fortune global 500 https://fortune.com/global500/search/

122. BBC Mozambique: Is Cabo Delgado the latest Islamic State outpost? https://www.bbc.com/news/world-africa-52532741
123. Aramco cyber attack https://money.cnn.com/2015/08/05/technology/aramco-hack/index.html
124. BBC NHS ransomware attack https://www.bbc.com/news/technology-41753022
125. Happy Planet Index http://happyplanetindex.org/about#how
126. West Africa Off Grid Solar https://www.pv-magazine.com/2019/04/24/off-grid-solar-in-africa-gets-224-million-world-bank-lift/
127. Documentary film 'American Factory' Netflix 2019 https://www.netflix.com/za/title/81090071
128. Exxonmobil climate lawsuit https://www.nytimes.com/2019/12/10/climate/exxon-climate-lawsuit-new-york.html
129. 4oceans  https://4ocean.com/
130. African entrepreneur collective https://africanentrepreneurcollective.org
131. Climate change could rain on Saudi Aramco's IPO parade Reuters https://uk.reuters.com/article/uk-aramco-ipo-esg-insight-idUKKCN1VA0EU
132. Germany wind and solar power WEF https://www.weforum.org/agenda/2020/08/where-solar-wind-power-are-thriving/#:~:text=Germany%20is%20well%20above%20the,through%20June%20of%20this%20year.&text=The%20world's%20strongest%20economies%20still,in%20China%2C%20India%20and%20Japan.
133. Money Logging Strauman L  https://www.money-logging.org/
134. Rewcastle Brown C. The Sarawak Report Gerakbudya Enterprise 2018 https://www.gerakbudaya.com/the-sarawak-report-the-inside-story-of-the-1mdb-expose
135. Shell widows lawsuit Wiwa https://www.reuters.com/article/us-shell-widows-lawsuit/dutch-court-to-hear-case-vs-shell-brought-by-widows-of-hanged-nigeria-activists-idUSKCN1S73CY
136. Governance definition https://www.governanceinstitute.com.au/resources/what-is-governance/
137. UN Universal Human Rights Index https://uhri.ohchr.org/en/sdgs
138. HPI eco-footprint Qatar https://www.dohanews.co/qatar-ranked-third-worst-in-happy-planet-index-for/

## Chapter 3: Learning from History

1. 'Gasland' documentary film IMDB 2010 https://www.imdb.com/title/tt1558250/
2. Montreal Protocol https://en.wikipedia.org/wiki/Montreal_Protocol
3. Tetra ethyl lead https://en.wikipedia.org/wiki/Tetraethyllead
4. IPCC Working Group 1 (WG1), 4th Assessment Report; 2007 https://www.ipcc.ch/report/ar4/wg1/
5. UNFCCC https://unfccc.int/
6. Hydrocarbon Processing Startups strive to recycle emissions for 'new carbon economy' 5/30/2019 https://www.hydrocarbonprocessing.com/news/2019/05/startups-strive-to-recycle-emissions-for-new-carbon-economy
7. Carson 'Chemistry That Kills and Rachel Carson – Why Silent Spring Says Don't Put DDT on Your Cerea'l utube https://www.youtube.com/watch?v=Oj5CjLNDr0o&frags=wn&ab_channel=MyGirlHeroes
8. Chernobyl New Safe Confinement wikipedia https://en.wikipedia.org/wiki/Chernobyl_New_Safe_Confinement
9. International Maritime Organization (IMO) http://www.imo.org/en/MediaCentre/PressBriefings/Pages/MEPC-70-2020sulphur.aspx
10. International Union of Conservation of Nature https://www.iucn.org/resources/issues-briefs/marine-plastics
11. Alliance for Plastic Waste https://endplasticwaste.org/
12. International Whaling Commission https://iwc.int/commercial
13. Reuters Japan Whaling https://www.reuters.com/article/us-japan-whaling/first-whales-caught-as-japan-resumes-commercial-hunt-after-30-years-idUSKCN1TW16U
14. Willie Smits https://en.wikipedia.org/wiki/Willie_Smits
15. New York Times Ethiopia https://www.nytimes.com/2019/07/30/world/africa/ethiopia-tree-planting-deforestation.html
16. Global Times China http://www.globaltimes.cn/content/1139006.shtml
17. Bangkok Post Thailand https://www.bangkokpost.com/thailand/general/1518826/new-laws-

chased-to-boost-afforestation

18. Time Magazine top ten in 2010 Kuwait oil fires
    http://content.time.com/time/specials/packages/article/0,28804,1986457_1986501_1986442,00.html
19. Exxon Valdez
    http://content.time.com/time/specials/packages/article/0,28804,1986457_1986501_1986446,00.html
20. Film 'Deepwater Horizon' IMDB 2016
    https://www.imdb.com/title/tt1860357/
21. Steffy LC Drowning in Oil Mc Graw Hill 2011
    https://www.amazon.com/Drowning-Oil-Reckless-Pursuit-Profit/dp/0071760814
22. The worst case of oil pollution on the planet Yasuni
    http://www.sosyasuni.org/en/index.php?option=com_content&view=article&id=106%3Athe-worst-case-of-oil-pollution-on-the-planet&catid=17%3Ageneral&Itemid=1
23. Documentary film 'Crude'
    https://www.youtube.com/watch?v=fB9jQryg6aQ
24. BBC Ecuador pollution case https://www.bbc.com/news/world-latin-america-45455984
25. Niger Delta pollution
    https://en.wikipedia.org/wiki/Environmental_issues_in_the_Niger_Delta
26. Time Magazine top ten in 2010 Bopal Disaster
    http://content.time.com/time/specials/packages/article/0,28804,1986457_1986501_1986445,00.html
27. Time Magazine top ten in 2010 Seveso disaster
    http://content.time.com/time/specials/packages/article/0,28804,1986457_1986501_1986449,00.html r
28. Chernobyl TV mini series IMDB 2019
    https://www.imdb.com/title/tt7366338/
29. Tokaimura A Brief History of: The Tokaimura Criticality Incident (Short Documentary) utube https://www.youtube.com/watch?v=acpz3CG1xi4
30. The Guardian Fukushima
    https://www.theguardian.com/environment/2019/sep/10/fukushima-japan-will-have-to-dump-radioactive-water-into-pacific-minister-says
31. Fukushima wikipedia

https://en.wikipedia.org/wiki/Fukushima_Daiichi_nuclear_disaster
32. Cape Industries lawsuit https://www.business-humanrights.org/en/capegencor-lawsuits-re-so-africa-0
33. Cape lawsuit https://www.business-humanrights.org/en
34. Asbestos law https://en.wikipedia.org/wiki/Asbestos_and_the_law
35. Whats where in the world Dorling Kindesley 2013 https://www.amazon.com/Whats-Where-World-DK/dp/1409379248
36. Health warning from Borneo fires https://en.wikipedia.org/wiki/2019_Southeast_Asian_haze
37. Straumann https://www.money-logging.org/
38. Borneo forestry map https://atlas.cifor.org/borneo/#en
39. Big 6 logging companies https://www.forbes.com/lists/2012/84/malaysia-billionaires-12_Yaw-Teck-Seng-Yaw-Chee-Ming_GSVV.html
40. Timber corruption https://www.stop-timber-corruption.org/
41. Liberia logging concessions https://www.globalwitness.org/en/archive/signingtheirlivesaway/
42. Sarawak report logging http://www.sarawakreport.org/2013/01/taib-family-logging-company-kicked-out-of-liberia/
43. Norwegian divestment https://www.culturalsurvival.org/news/norwegian-government-declares-malaysian-timber-giant-unethical-company
44. Time Magazine top ten in 2010 Love Canal http://content.time.com/time/specials/packages/article/0,28804,1986457_1986501_1986441,00.html
45. Hinkley and PG&EC Wikipedia https://en.wikipedia.org/wiki/Hinkley_groundwater_contamination
46. Documentary film 'Erin Brockovich' IMDB 2000 https://www.imdb.com/title/tt0195685/
47. Documentary film 'The Devil We Know' Dupont class action 2018 https://www.imdb.com/title/tt7689910/
48. CNN Flint water https://edition.cnn.com/2016/03/04/us/flint-water-crisis-fast-facts/index.html
49. National Geographic Flint water crisis https://www.nationalgeographic.com/environment/2019/04/flint-water-crisis-fifth-anniversary-flint-river-pollution/

50. Documentary film 'Minamata: The victims and their world' IMDB 1971 https://www.imdb.com/title/tt0068951/
51. Maldives corruption https://www.occrp.org/en/paradiseleased/how-paradise-was-carved-up-and-sold
52. Maldives ex president conviction for corruption https://www.bbc.com/news/world-asia-50590921
53. Encyclopedia Britannica Equatorial Guinea https://www.britannica.com/place/Equatorial-Guinea/Independence#ref517150
54. Asean flu Wikipedia https://en.wikipedia.org/wiki/1997_Asian_financial_crisis
55. Dot-com crash Wikipedia https://en.wikipedia.org/wiki/Dot-com_bubble
56. Enron Wikipedia https://en.wikipedia.org/wiki/Enron_scandal
57. SOX https://www.thebalancesmb.com/sarbanes-oxley-act-and-the-enron-scandal-393497
58. Financial Crisis Inquiry Commission – Press Release – January 27, 2011 https://www.loc.gov/item/2011381760/
59. McLean B. and Nocera J. All the Devils are here - The hidden history of the financial crisis Portfolio Penguin 2010 https://www.amazon.com/All-Devils-Are-Here-Financial/dp/159184438X
60. Tedtalks Lucy Koechin Power of Corruption https://www.youtube.com/watch?v=295T3fUF1SM
61. Tedtalks Sarah Chayes How Government Corruption is a Precursor to Extremism https://tedxfultonstreet.com/people/sarah-chayes/
62. Transparency International global analysis https://www.transparency.org/news/feature/cpi_2018_global_analysis
63. OECD Anti Bribery Commission https://www.oecd.org/corruption-integrity/
64. Shaik corruption trial South Africa https://en.wikipedia.org/wiki/Schabir_Shaik_trial
65. South China Morning Post Malaysia corruption https://www.scmp.com/news/asia/southeast-asia/article/2174173/malaysia-reopens-investigation-french-submarine-deal
66. Thales slush fund https://en.wikipedia.org/wiki/Thales_Group

67. Guardian Siemens slush fund https://www.theguardian.com/technology/2008/jan/25/1
68. Rewcastle Brown C. The Sarawak Report Gerakbudya Enterprise 2018 https://www.gerakbudaya.com/the-sarawak-report-the-inside-story-of-the-1mdb-expose
69. Ukio bankas money laundering https://emerging-europe.com/news/lithuanias-now-defunct-ukio-bankas-was-used-as-laundromat-report-claims/
70. Venezuela in crisis https://www.aljazeera.com/news/2019/01/venezuela-crisis-latest-updates-190123205835912.html
71. 2022 FIFA world cup https://en.wikipedia.org/wiki/Qatar_2022_FIFA_World_Cup_bid
72. Trioka Laundromat occrp https://www.occrp.org/en/troikalaundromat/
73. Al Jazeera Maldives money laundering https://www.aljazeera.com/investigations/stealing-paradise/
74. IOL Corporate Corruption Report: SA milked billions in corrupt deals https://www.iol.co.za/business-report/corporatecorruptionreport-sa-milked-billions-in-corrupt-deals-10284489
75. ICIJ Panama Paper https://www.icij.org/investigations/panama-papers/pages/panama-papers-about-the-investigation/
76. ICIJ Paradise papers https://www.icij.org/tags/glencore/
77. The Guardian Gurtel case https://www.theguardian.com/news/2019/mar/01/spain-watergate-corruption-scandal-politics-gurtel-case
78. SEC press release Hitachi https://www.sec.gov/news/pressrelease/2015-212.html
79. BBC news Odebrecht corruption scandal https://www.bbc.com/news/world-latin-america-41109132
80. BBC news Odebrecht corruption scandal https://www.bbc.com/news/business-39194395
81. Jade and Gem Smuggling in MyanmarThe Diplomat https://thediplomat.com/2019/03/jade-and-gem-smuggling-costs-myanmar-billions/
82. Glencore, Gertier and Kabila https://www.hrw.org/world-report/2019/country-chapters/democratic-republic-congo
83. Glencore and Glasenberg https://www.businesslive.co.za/bd/world/2018-

05-21-dan-gertler-bribe-allegations-back-to-haunt-glencores-ivan-glasenberg/
84. Nigeria Abacha BBC https://www.bbc.co.uk/news/resources/idt-f9f1cd17-2c50-442e-88fc-e2deb46dbde1
85. Peru Fujimori https://www.bbc.com/news/world-latin-america-16097439
86. Tunisia Ben Ali BBC https://www.bbc.com/news/world-africa-49752876
87. Ukraine The Guardian https://www.theguardian.com/world/2019/jan/25/ukraine-ex-president-viktor-yanukovych-found-guilty-of-treason
88. Supercars Equatorial Guinea corruption https://www.reuters.com/article/us-swiss-equatorial-cars/supercars-seized-from-equatorial-guinea-vice-president-net-27-million-at-auction-idUSKBN1WE0NV
89. Al Jazeera The Gambia https://www.aljazeera.com/news/2019/03/gambia-president-yahya-jammeh-stole-362m-190329181812901.html
90. Carson R Silent Spring Penguin 1962 https://www.amazon.com/Silent-Spring-Penguin-Modern-Classics/dp/0141184949
91. Polychlorinated biphenyls (PCBs) https://www.center4research.org/pcbs-still-causing-harm-decades-ban/
92. World Bank sanction report https://www.worldbank.org/en/news/press-release/2018/10/03/world-bank-group-debarred-78-firms-and-individuals-during-fiscal-year-2018
93. Minamata 60 years later https://www.japantimes.co.jp/opinion/2017/01/21/editorials/relief-minamata-victims/
94. Tar sands of Canada https://www.nationalgeographic.com/environment/2019/04/alberta-canadas-tar-sands-is-growing-but-indigenous-people-fight-back/
95. Time Magazine top ten in 2010 Aral Sea http://content.time.com/time/specials/packages/article/0,28804,1986457_1986501_1986451,00.html
96. NASA report Ozone hole https://www.nasa.gov/feature/goddard/2019/2019-ozone-hole-is-the-smallest-on-record-since-its-discovery
97. Documentary film 'How to steal a country' daily motion

https://www.dailymotion.com/video/x7tumo6

98. Documentary film 'The Edge of Democracy' IMDB 2019
https://www.imdb.com/title/tt06016744/

99. Corporate Finance Institute Top Accounting Scandals
https://corporatefinanceinstitute.com/resources/knowledge/other/top-accounting-scandals/

100. Enron scandal https://en.wikipedia.org/wiki/Enron_scandal

101. Benzinga Arthur Anderson
https://www.benzinga.com/markets/14/04/4429482/how-the-arthur-anderson-and-enron-fraud-changed-accounting-forever

102. The Telegraph Arthur Andersen returns 12 years after Enron scandal
https://www.telegraph.co.uk/finance/newsbysector/banksandfinance/11069713/Arthur-Andersen-returns-12-years-after-Enron-scandal.html#:~:text=One%20of%20the%20financial%20world's,accounting%20scandal%20saw%20it%20vanish.&text=They%20will%20rename%20their%20San,respected%20in%20US%20financial%20history.

103. KPMG https://home.kpmg/xx/en/home/about/who-we-are/governance.html

104. This is Money Deloitte fined
https://www.thisismoney.co.uk/money/markets/article-6650093/Beancounter-Deloitte-fined-Malaysia-sovereign-wealth-fund-scandal.html

105. Rewcastle Brown C The Sarawak Report pages 222 KPMG and 389 Banks
https://www.amazon.com/Sarawak-Report-Inside-Story-Expos%C3%A9/dp/1527219364

106. 'Active Measures' documentary film Bryan J IMDB 2018
https://www.imdb.com/video/vi860797465?playlistId=tt8135494&ref_=tt_ov_vi

107. Noseweek magazine page 27 September 2018
https://www.noseweek.co.za/archives/227/September-2018

108. Al Araby Libya Sovereign Wealth Fund
https://www.alaraby.co.uk/english/indepth/2017/11/30/libyas-sovereign-wealth-scandal-taxpayers-billions-squandered-through-incompetence

109. Finestein A Shadow World Documentary Film
https://www.amazon.com/Shadow-World-Johan-Grimonprez/dp/B01M29SMYE

## Chapter 4: Measuring Sustainable Success

1. Senge P etc. The Fifth Discipline Fieldbook: Strategies for Building a Learning Organization 1990 https://www.amazon.co.uk/Fifth-Discipline-Fieldbook-Strategies-Organization/dp/1857880609
2. Fragile States Index https://fundforpeace.org/2019/04/10/fragile-states-index-2019/
3. Poorest Countries Index https://www.focus-economics.com/blog/the-poorest-countries-in-the-world
4. Misery Index https://www.forbes.com/sites/stevehanke/2019/03/28/hankes-annual-misery-index-2018-the-worlds-saddest-and-happiest-countries/#2d49b16f3bce
5. Corruption Perception Index https://www.transparency.org/en/cpi/2019/results/
6. Bribe payers index https://www.transparency.org/whatwedo/publication/bpi_2011
7. Anti Bribery and Corruption Benchmarking Report https://img04.en25.com/Web/DuffPhelps/%7Bb873e07b-a163-4ceb-80fa-5c01c8ee5a7f%7D_anti-bribery-and-corruption-benchmarking-report-2018.pdf
8. Fraud and Risk Report https://img04.en25.com/Web/DuffPhelps/%7Be9591fd6-feb1-4d8f-889f-a49650fcc564%7D_global-fraud-and-risk-report-17.pdf
9. Nigeria SEC https://www.petroleum-economist.com/articles/politics-economics/africa/2019/oando-under-fire-from-nigerias-sec
10. Money Laundering Index https://www.baselgovernance.org/asset-recovery/basel-aml-index
11. Financial Action Task Force (FATF) http://www.fatf-gafi.org/publications/mutualevaluations/?hf=10&b=0&s=desc(fatf_releasedate)
12. Moodys https://www.moodys.com/
13. S&P https://www.spglobal.com/en/capabilities/credit-ratings
14. SP global https://www.spglobal.com/en/capabilities/esg-evaluation
15. Fitch https://www.fitchratings.com/site/definitions
16. Dagong

http://www.dagonghk.com/Procedures.php?act=list&parent_id=54&menu_id=870
17. Ethisphere ethical companies https://www.worldsmostethicalcompanies.com/honorees/
18. Sustainability https://www.robecosam.com/csa/indices/?r
19. Dow Jones Sustainability Index (DJSI) leaders https://www.spglobal.com/esg/csa/csa-resources/
20. Sustainability indices https://www.sustainability-indices.com/
21. Baldridge Award https://en.wikipedia.org/wiki/Malcolm_Baldrige_National_Quality_Award
22. EFQM https://www.bqf.org.uk/what-we-do/performance-improvement/efqm-excellence-model/
23. Australian Business Excellence Awards https://businessexcellenceawards.com.au/
24. ASIAN www.acga-asia.org
25. CIA GDP https://www.cia.gov/library/publications/the-world-factbook/fields/208rank.html
26. Social Progress Index https://www.socialprogress.org/assets/downloads/resources/2019/2019-Social-Progress-Index-executive-summary-v2.0.pdf
27. Fortunes 50 most admired companies http://fortune.com/worlds-most-admired-companies/
28. IMD World Competitiveness Ranking and World Talent Ranking www.imd.org
29. Equator principles https://equator-principles.com/
30. Organization for Economic Cooperative Development (OECD) https://stats.oecd.org/Index.aspx?DataSetCode=PDB_GR
31. Ibrahim governance index http://mo.ibrahim.foundation/iiag/
32. Freedom House https://freedomhouse.org/
33. The Guardian carbon emissions https://www.theguardian.com/environment/2019/oct/09/revealed-20-firms-third-carbon-emissions
34. Straits Times BP climate change https://www.straitstimes.com/world/europe/bps-new-ceo-bernard-looney-goes-all-in-on-climate-goals-and-explores-overhaul

35. Shell https://www.shell.com/sustainability/our-approach/un-sustainable-development-goals.html#iframe=L3dlYmFwcHMvc2hlbGwtc2RnLw .
36. Task Force on Climate-related Financial Disclosures: Status Report 2: 2019 https://www.fsb.org/2019/06/task-force-on-climate-related-financial-disclosures-2019-status-report/
37. Fairtrade coffee https://www.fairtrade.net/
38. Fairtrade promotion http://www.fairtrade.org.uk/What-is-Fairtrade/What-Fairtrade-does
39. Woolworths South Africa https://www.woolworthsholdings.co.za/wp-content/uploads/2019/09/WHL_GBJ_Abridged_Report.pdf
40. Tesla ceo and chairman Musk https://www.scmp.com/news/world/united-states-canada/article/2166334/elon-musk-removed-tesla-chairman-will-remain-ceo
41. Blackrock https://www.eco-business.com/opinion/blackrock-is-getting-serious-about-climate-change-is-this-a-turning-point-for-investors/
42. Ikea https://newsroom.inter.ikea.com/news/ikea-sustainability-report-fy18/s/409b1d78-7d7b-4f48-a63d-368d5fb69c2d
43. Kimberly Clark https://www.kimberly-clark.com/en/responsibility/sustainability2022
44. Kimberly Clark sustainability https://www.pulpandpapercanada.com/kimberly-clark-progressing-toward-sustainability-goals-1100001239/
45. Shangri la hotels and resorts https://www.shangri-la-sustainability.com/our-environment
46. Climate progress tracker energy https://unclimatesummit.org/trackerhome/trackerenergy/
47. Climate progress tracker companies https://unclimatesummit.org/trackerhome/trackerbusiness/
48. Global Board Diversity Tracker 2018: Who's Really On Board? Egon Zehnder https://www.egonzehnder.com/global-board-diversity-tracker/take-action/unilever-ceo-paul-polman-on-diversitys-strategic-importance#
49. BP Statistical Review of World Energy 2019 68th edition https://www.bp.com/en/global/corporate/news-and-insights/press-releases/bp-statistical-review-of-world-energy-2019.html

50. Woolworths carbon footprint https://www.woolworthsholdings.co.za/wp-content/uploads/2019/09/2018-Carbon-Footprint-Overview-.pdf
51. FT IFC awards https://live.ft.com/Events/FT-IFC-Transformational-Business-Awards-2019
52. Euromoney bank awards https://www.euromoney.com/research-and-awards/surveys-and-awards/awards-for-excellence/2019
53. Harvard law board diversity https://corpgov.law.harvard.edu/2019/02/05/missing-pieces-report-the-2018-board-diversity-census-of-women-and-minorities-on-fortune-500-boards/
54. WSJ chairman and ceo roles https://www.wsj.com/articles/more-u-s-companies-separating-chief-executive-and-chairman-roles-11548288502
55. EU Competition Commission https://ec.europa.eu/competition/
56. Transparency International (TI) www.transparency.org
57. Kroll Global fraud https://www.kroll.com/en/insights/publications/global-fraud-and-risk-report-2019
58. Al Jazeera Fudai wall https://www.youtube.com/watch?v=0LLD3Ww4V40
59. UNIPCC https://www.ipcc.ch/
60. Gladwell M The Tipping Point Little, Brown and Company 2000 https://www.amazon.com/Tipping-Point-Little-Things-Difference/dp/0316346624
61. The Guardian Superbug hot spots https://www.theguardian.com/environment/2019/sep/19/superbug-hotspots-emerging-in-farms-across-globe-study?CMP=Share_iOSApp_Other
62. DNVGL Energy Transion Outlook 2020 https://eto.dnvgl.com/2020/index.html
63. Renault CEO and Chairman https://www.wsj.com/articles/splitting-ceo-chairman-roles-expected-to-improve-renaults-governance-1542738407

## Chapter 5: Global Warming and State of Earth

1. World Bank press release https://www.worldbank.org/en/news/press-release/2018/03/19/climate-change-could-force-over-140-million-to-migrate-within-countries-by-2050-world-bank-report
2. UN report https://www.theguardian.com/environment/2019/may/06/human-society-under-urgent-threat-loss-earth-natural-life-un-report
3. United Nations Framework Convention on Climate Change (UNFCCC) https://en.wikipedia.org/wiki/2016_United_Nations_Climate_Change_Conference
4. World Economic Forum https://www.weforum.org/agenda/2019/06/chart-of-the-day-these-countries-create-most-of-the-world-s-co2-emissions/
5. Documentary film 'Chasing Coral' 2017 Netflix https://www.chasingcoral.com/
6. Dorling Kindesley 'Whats Where in the World' https://www.dk.com/uk/book/9781409379249-whats-where-in-the-world/
7. Documentary film 'Chasing Ice' IMDB 2012 https://www.imdb.com/title/tt1579361/
8. Alaska permafrost: Anchorage Daily News https://www.adn.com/alaska-news/weather/2019/12/17/as-alaska-permafrost-melts-roads-sink-bridges-tilt-and-greenhouse-gases-escape/#:~:text=By%20the%20century's%20end%2C%20even,University%20of%20Alaska%20Fairbanks%20research.&text=Oil%
9. Siberian forest fires Mongabay https://news.mongabay.com/2020/07/photos-show-scale-of-massive-fires-tearing-through-siberian-forests/#:~:text=A%20series%20of%20newly%20released,since%20the%20start%20of%202020.
10. UK peatbogs The Independent https://www.independent.co.uk/news/uk/peat-bog-destruction-boosts-global-warming-1540458.html
11. Borneo forest fires NASA https://earthobservatory.nasa.gov/images/145614/smoke-blankets-borneo#:~:text=After%20several%20relatively%20quiet%20fire,pall%20of

%20thick%2C%20noxious%20smoke.&text=Smoke%20hovered%20over%20the%20islands,health%20warnings%20in%20the%20region.
12. Mangrove swamps Mongabay https://news.mongabay.com/2018/07/why-mangroves-matter-experts-respond-on-international-mangrove-day/
13. Snow reflection National Geographic https://www.nationalgeographic.com/news/2014/6/140610-connecting-dots-dust-soot-snow-ice-climate-change-dimick/
14. Antibiotics WHO https://www.who.int/news-room/fact-sheets/detail/antimicrobial-resistance
15. Ebola WHO https://www.who.int/news-room/fact-sheets/detail/ebola-virus-disease
16. Hey RB Performance Management for the Oil, Gas and Process Industries Elsevier 2017 https://www.elsevier.com/books/performance-management-for-the-oil-gas-and-process-industries/hey/978-0-12-810446-0
17. Methane https://globalmethane.org/
18. Eskom Medupi project https://mg.co.za/article/2019-08-19-00-cancel-eskoms-odious-debt-to-the-world-bank
19. Sarawak Report http://www.sarawakreport.org/2012/02/scandal-of-salco-how-taib-plans-to-make-billions-from-bakun/
20. Science Based Targets https://www.energypolicy.solutions/guide/
21. BBC palm oil https://www.bbc.co.uk/newsround/39492207
22. LA Times Land use policy key to reining in global warming, U.N. report warns Rosen J and Phillips A Aug. 8, 2019 https://www.latimes.com/environment/story/2019-08-08/ipcc-land-use-global-warming
23. Hydrocarbon Processing plastic pollution https://www.hydrocarbonprocessing.com/news/2019/06/g20-to-tackle-ocean-plastic-waste-as-petrochemical-producers-expand
24. Science Advances https://advances.sciencemag.org/
25. Conway EM and Oreskes N 'Merchants of Doubt' 2010 https://www.amazon.com/Merchants-Doubt-Handful-Scientists-Obscured/dp/1608193942
26. Al Gore: An Inconvenient Truth documentary film https://www.imdb.com/title/tt0497116/

27. Paris Agreement https://en.wikipedia.org/wiki/Paris_Agreement
28. Trump climate agreement withdrawal https://www.imd.org/research-knowledge/articles/what-does-the-trump-administration-and-climate-change-denial-have-to-do-with-switzerland/
29. UNFCCC https://en.wikipedia.org/wiki/2016_United_Nations_Climate_Change_Conference
30. Hydrocarbon Processing G20 countries plastic waste https://www.hydrocarbonprocessing.com/news/2019/06/g20-to-tackle-ocean-plastic-waste-as-petrochemical-producers-expand
31. Science Magazine plastic trash https://www.sciencemag.org/tags/plastic-pollution
32. UN Climate Change Summit CAS 2019 https://www.un.org/en/climatechange/
33. COP25 2019 Madrid https://unclimatesummit.org/cop25/?gclid=EAIaIQobChMI29qqtuym6wIVU-3tCh0gLwnwEAAYASAAEgJ0APD_BwE
34. Guardian Mon 8 Oct 2018 We have 12 years to limit climate change catastrophe, warns UN https://www.theguardian.com/environment/2018/oct/08/global-warming-must-not-exceed-15c-warns-landmark-un-report
35. Intergovernmental Panel on Climate Change (IPCC) https://www.ipcc.ch/
36. CNN biodiversity report https://edition.cnn.com/2019/05/06/world/un-biodiversity-report-end-of-nature-sutter-scn/index.html
37. IPBES https://www.ipbes.net/
38. The Guardian biodiversity report https://www.theguardian.com/environment/2019/may/06/human-society-under-urgent-threat-loss-earth-natural-life-un-report
39. Fourth National Climate Assessment (NCA4) https://en.wikipedia.org/wiki/Fourth_National_Climate_Assessment
40. New York Times 'The Daily': The White House Plan to Change Climate Science https://www.nytimes.com/2019/05/29/podcasts/the-daily/trump-climate-science.html
41. 2019 BP Statistical Review of World energy https://www.bp.com/en/global/corporate/energy-economics/statistical-review-of-world-energy.html

42. UN Climate Action Summit 2019 https://unclimatesummit.org/?gclid=CjwKCAjw2qHsBRAGEiwAMbPoDLgFftWLY6L5YFe8ilvV8exlbwga3ywEbyAJEFVICTXKtSe78qQcfxoCr0UQAvD_BwE
43. Energy policy solutions https://www.energypolicy.solutions/guide/
44. Neste renewables https://www.hydrocarbonprocessing.com/news/2019/06/irpc-eurasia-19-neste-ceo-touts-energy-transformation-through-renewables-sustainability
45. US coal cost crossover https://energyinnovation.org/publication/the-coal-cost-crossover/
46. Europe's power industry CO2 costs https://www.petroleum-economist.com/articles/markets/trends/2019/carbon-price-drives-generating-fuel-switch
47. The climate group https://www.theclimategroup.org/
48. RE100 http://re100.org/
49. AEPW https://endplasticwaste.org/
50. AEPW CEO https://www.hydrocarbonprocessing.com/news/2019/08/alliance-to-end-plastic-waste-appoints-president-ceo
51. OGCI https://oilandgasclimateinitiative.com/
52. World Bank projects: regional electric power https://www.worldbank.org/en/topic/energy
53. NY Times Ethiopia tree planting record https://www.nytimes.com/2019/07/30/world/africa/ethiopia-tree-planting-deforestation.html
54. South Africa concentrated solar power https://www.ee.co.za/article/power-from-the-sun-an-overview-of-csp-in-south-africa.html
55. South Africa wind power https://en.wikipedia.org/wiki/List_of_wind_farms_in_South_Africa
56. South Africa solar PV power https://www.youtube.com/watch?v=Tq14KORM2Us&gl=UG&hl=en-GB
57. Butan Carbon Neutral Tedtalk https://www.ted.com/talks/tshering_tobgay_this_country_isn_t_just_carbon_neutral_it_s_carbon_negative?language=en
58. Australia renewables Petroleum Economics https://www.petroleum-

economist.com/articles/politics-economics/asia-pacific/2019/big-names-target-australian-renewables

59. Rystad energy https://www.rystadenergy.com/
60. Adelaide battery Musk https://www.bloomberg.com/news/articles/2020-02-28/two-years-on-musk-s-big-battery-bet-is-paying-off-in-australia
61. 'Willie Smits restores a rainforest' Tedtalks https://www.ted.com/talks/willie_smits_restores_a_rainforest?language=en
62. Business Round Table https://www.businessroundtable.org/
63. Global Sustainable Investment Review http://www.gsi-alliance.org/wp-content/uploads/2019/06/GSIR_Review2018F.pdf,
64. Institutional divestment https://www.energypolicy.solutions/guide/
65. Exxon and Shell divestment https://www.hydrocarbonprocessing.com/news/2019/08/climate-change-could-rain-on-saudi-aramcos-ipo-parade
66. BP investments Hydrocarbon Processing https://www.hydrocarbonprocessing.com/news/2019/06/breakthrough-technology-to-produce-animal-feeds-from-natural-gas
67. BP and Bunge Hydrocarbon Processing https://www.hydrocarbonprocessing.com/news/2019/07/major-renewable-energy-expansion-to-create-world-class-bioenergy-company
68. Low carbon energy Petroleum Economics 17 July 2019 https://www.petroleum-economist.com/articles/low-carbon-energy/renewables/2019/the-climate-threat-will-investors-exit-big-oil
69. Accoina https://www.acciona.com/
70. Mainstream https://www.mainstreamrp.com/
71. North Sea Wind Farms https://northseawindpowerhub.eu/key-players-wind-industry-support-ex-amination-feasability-of-north-sea-wind-power-hub/
72. Storing solar energy in the strangest places: Will Chueh at TEDxStanford https://www.youtube.com/watch?v=aFaOr05gvpI
73. Total biofuels refinery Hydrocarbon Processing https://www.hydrocarbonprocessing.com/news/2019/07/total-starts-biofuel-production-at-la-mede-refinery
74. Total biofuels refinery Hydrocarbon Processing https://www.hydrocarbonprocessing.com/news/2019/05/total-to-move-

ahead-with-using-palm-oil-at-biodiesel-refinery
75. Hydrocarbon Processing Neste MY Renewable diesel launched in Sweden 11/7/2018
https://www.hydrocarbonprocessing.com/news/2018/11/neste-my-renewable-diesel-launched-in-sweden
76. SkyNRG aviation fuel Hydrocarbon Processing
https://www.hydrocarbonprocessing.com/news/2019/05/skynrg-chooses-topsoe-technology-for-europe-s-first-sustainable-aviation-fuel-plant
77. Hydrocarbon Processing Clariant and Neste join forces to develop sustainable industrial solutions 11/6/2018
https://www.hydrocarbonprocessing.com/news/2018/11/clariant-and-neste-join-forces-to-develop-sustainable-industrial-solutions
78. Hydrocarbon Processing EM lower emissions ExxonMobil to invest up to $100M on lower-emissions R&D 5/8/2019
https://www.hydrocarbonprocessing.com/news/2019/05/exxonmobil-to-invest-up-to-100m-on-lower-emissions-rd
79. Hydrocarbon Processing CO2 Startups strive to recycle emissions for 'new carbon economy' 5/30/2019
https://www.hydrocarbonprocessing.com/news/2019/05/startups-strive-to-recycle-emissions-for-new-carbon-economy
80. Lanzatech Jingtang Steel https://www.lanzatech.com/2018/06/08/worlds-first-commercial-waste-gas-ethanol-plant-starts/
81. Ink from diesel exhausts Fortune Magazine
https://fortune.com/2017/02/12/grviky-labs-india-car-exaust-ink/
82. IEA World Energy Outlook 2019 Dr Fatih Birol, IEA Executive Director
https://www.iea.org/reports/world-energy-outlook-2019
83. Canada's Changing Climate Report CCCR 2019.
https://changingclimate.ca/cccr2019/
84. Global Methane Initiative
https://www.globalmethane.org/about/index.aspx
85. Borneo project https://borneoproject.org/updates/bmf-press-release-malaysian-lawyers-slam-sarawak-dam-resettlements
86. Mongabay editor arrest Asia Sentinal
https://www.asiasentinel.com/p/concern-grows-in-indonesia-over-environmentalist

87. Green Climate Fund (GCF) https://www.greenclimate.fund/home
88. International Solar Alliance (ISA) https://isolaralliance.org/
89. Solar power in California https://en.wikipedia.org/wiki/Solar_power_in_California
90. Solar power in Arizona https://en.wikipedia.org/wiki/Solar_power_in_Arizona
91. Documentary film 'Happening: A Clean Energy Revolution.' IMDB 2017 https://www.imdb.com/title/tt7212266/
92. Solar power in Nevada https://en.wikipedia.org/wiki/Solar_power_in_Nevada
93. Ethanol fuel in Brazil https://en.wikipedia.org/wiki/Ethanol_fuel_in_Brazil
94. PV magazine South Africa Solar power https://www.pv-magazine.com/2019/10/21/south-africa-plans-to-allocate-at-least-6-gw-of-large-scale-solar-by-2030/
95. Capital costs Petroleum Economics James Mills https://www.petroleum-economist.com/articles/corporate/finance/2019/capital-costs-rise-on-sustainability-concerns
96. BP and the net zero Teesside project https://www.bp.com/en/global/corporate/news-and-insights/bp-magazine/net-zero-teesside-project.html
97. Ballard fuel cells https://www.ballard.com/
98. France biofuels palm oil ruling Hydrocarbon Processing https://www.hydrocarbonprocessing.com/news/2019/10/in-blow-to-total-france-upholds-law-banning-palm-oil-from-biofuel-scheme
99. Neste sustainable diesel Hydrocarbon Processing https://www.hydrocarbonprocessing.com/news/2019/10/neste-my-renewable-diesel-now-available-in-all-baltic-countries-including-estonia?id=3872815
100. Neste collaboration Hydrocarbon Processing https://www.hydrocarbonprocessing.com/news/2019/10/neste-collaborates-for-a-more-sustainable-aviation?id=3872815
101. EWE Hydrogen Hydrocarbon Processing https://www.hydrocarbonprocessing.com/news/2020/02/germanys-ewe-to-test-supplying-hydrogen-for-transport-industry
102. Borealis renewable polypropylene Hydrocarbon Processing 16October

2019 https://www.hydrocarbonprocessing.com/news/2019/10/borealis-to-produce-renewable-polypropylene-using-neste-s-renewable-propane?id=3872815
103. The Balance Plastic recycling https://www.thebalancesmb.com/an-overview-of-plastic-recycling-4018761
104. Biggest companies in the world Fortune global 500 https://fortune.com/global500/search/
105. EU Green Deal Alessandro Vitelli 28 February 2020 Petroleum Econ https://www.petroleum-economist.com/articles/low-carbon-energy/energy-transition/2020/eu-s-green-deal-threatens-seismic-shift
106. Petroleum Economist Investors https://www.petroleum-economist.com/articles/low-carbon-energy/energy-transition/2020/investors-and-ngos-take-tougher-line-on-climate-change
107. Hydrocarbon Processing Shell and Gasunie plan to build massive Dutch green hydrogen plant 2/27/2020 https://www.hydrocarbonprocessing.com/news/2020/02/shell-and-gasunie-plan-to-build-massive-dutch-green-hydrogen-plant
108. Diamond green diesel Hydrocarbon Processing https://www.hydrocarbonprocessing.com/news/2019/09/louisiana-facility-plans-2021-expansion-to-meet-growing-demand-for-renewable-fuels
109. DOW UPM Finland Hydrocarbon Processing https://www.hydrocarbonprocessing.com/news/2019/09/dow-and-upm-partner-to-produce-plastics-made-with-renewable-feedstock
110. Repsol circular polyolefins Hydrocarbon Processing Repsol certifies petrochemical complexes for the production of circular polyolefins 2/28/2020 https://www.hydrocarbonprocessing.com/news/2020/02/repsol-certifies-petrochemical-complexes-for-the-production-of-circular-polyolefins?id=3872815
111. Lanzatech https://www.lanzatech.com/2020/06/02/lanzajet-takes-off/
112. Hydrocarbon Processing Saudi hydrogen project https://www.hydrocarbonprocessing.com/news/2020/07/world-s-largest-green-hydrogen-project-will-use-haldor-topsoe-ammonia-technology?id=3872815
113. Let the environment guide our development | Johan Rockstrom https://www.youtube.com/watch?v=RgqtrlixYR4&ab_channel=TED

114. Coral reef crisis guide https://sloactive.com/coral-reef-crisis-guide/
115. World Health Organization (WHO) https://www.who.int/foodsafety/areas_work/food-technology/faq-genetically-modified-food/en/
116. Cow manure energy South Africa https://www.youtube.com/watch?v=2-QruMN7wTA&ab_channel=AFD-AgenceFran%C3%A7aisedeD%C3%A9veloppement
117. Documentary film 'Merchants of Doubt' IMDB 2014 https://www.imdb.com/title/tt3675568/
118. Documentary film Al Gore An Inconvenient Sequel https://www.imdb.com/title/tt6322922/
119. UN Summit on Biodiversity https://www.un.org/pga/74/united-nations-summit-on-biodiversity/
120. Leaders pledge on biodiversity https://www.leaderspledgefornature.org/
121. ExxonMobil Global Therostat agreement https://corporate.exxonmobil.com/News/Newsroom/News-releases/2020/0921_ExxonMobil-expands-agreement-with-Global-Thermostat-re-direct-air-capture-technology
122. VW recycling https://www.volkswagenag.com/en/sustainability/environment/recycling.html
123. Mckinsey https://www.mckinsey.com/industries/chemicals/our-insights/how-plastics-waste-recycling-could-transform-the-chemical-industry#
124. 'Why we are storing billions of seeds' Jonathan Drori Tedtalk https://www.ted.com/talks/jonathan_drori_why_we_re_storing_billions_of_seeds#t-376992
125. 'The Other Inconvenient Truth' Jonathan Foley Tedtalk https://www.ted.com/talks/jonathan_foley_the_other_inconvenient_truth
126. China commitment to carbon neutrality by 2060 https://www.climatechangenews.com/2020/09/22/xi-jinping-china-will-achieve-carbon-neutrality-2060/
127. Mongabay https://news.mongabay.com/
128. Kiss the Ground Documentary film 2020 Netflix https://www.netflix.com/title/81321999

129. NYT Roundup Maker to Pay $10 Billion to Settle Cancer Suits https://www.nytimes.com/2020/06/24/business/roundup-settlement-lawsuits.html
130. A Plastic Wave Documentary Film https://www.youtube.com/watch?v=9-dpv2xbFyk
131. IPCC Report on Global Warming https://www.ipcc.ch/2018/10/08/summary-for-policymakers-of-ipcc-special-report-on-global-warming-of-1-5c-approved-by-governments/
132. Big Oil and Global Warming https://insideclimatenews.org/news/15072020/oil-gas-climate-pledges-bp-shell-exxon
133. Carbon Disclosure Project (CDP) https://www.cdp.net/en

## Chapter 6: The Future: - What can be done

1. World Bank 'Eliminating extreme poverty by 2030' https://www.youtube.com/watch?v=K7Mu72mZbSU .
2. Bob Dudley Group chief executive BP June 2019 https://www.bp.com/en/global/corporate/news-and-insights/press-releases/bp-statistical-review-of-world-energy-2019.html
3. The Guardian Earth Report https://www.theguardian.com/environment/2019/may/06/human-society-under-urgent-threat-loss-earth-natural-life-un-report
4. $CO_2$ emissions source: World Resources Institute.https://www.wri.org/
5. Breaking the Mould: The Role of Scenarios in Shaping South Africa's Future Segal N SUN press 2007 https://www.amazon.com/Breaking-Mould-Scenarios-Shaping-Africas/dp/1920109927
6. Reospartners scenario planning https://reospartners.com/learning-from-experience-the-mont-fleur-scenario-exercise/
7. Shell energy scenarios to 2050 https://www.youtube.com/watch?v=jQ2uIPeiEYQ
8. Shell scenarios https://www.shell.com/energy-and-innovation/the-energy-future/scenarios/shell-scenarios-in-film.html
9. Climate Interactive and Climate Action Tracker.
10. Exxonmobil carbon capture https://www.hydrocarbonprocessing.com/news/2019/08/exxonmobil-to-explore-new-carbon-capture-technology
11. Investment in clean power https://www.petroleum-economist.com/articles/low-carbon-energy/renewables/2019/emerging-market-clean-power-offers-investment-option
12. Power https://www.ucsusa.org/resources/environmental-impacts-natural-gas
13. Hydrocarbon Processing Biofuels https://www.hydrocarbonprocessing.com/news/2019/08/biofuels-company-proposes-to-buy-fire-damaged-refinery
14. Cleantechnica EVs https://cleantechnica.com/2019/08/12/bnp-parabas-says-the-party-is-over-for-oil-companies/
15. Petroleum Economist China evs https://www.petroleum-economist.com/articles/midstream-downstream/vehicles/2019/china-

spearheading-ev-adoption
16. Financial Times prize Lithium EVs https://live.ft.com/Events/FT-IFC-Transformational-Business-Awards-2019
17. BP World Energy Review 2018 carbon footprint page 12
18. Fight Cancer Network US https://www.fightcancer.org/news/department-justice-lawsuit-against-tobacco-industry
19. Canada tobacco companies https://www.business-humanrights.org/en/canada-tobacco-companies-lose-appeal-in-class-action-lawsuit-by-smokers-about-lack-of-warning-re-risks-of-smoking-ordered-to-pay-ca17-billion-in-damages
20. Documentary film 'AlphaGo' IMDB 2017 https://www.imdb.com/title/tt6700846/
21. X company www.x.company
22. Lanzatech Jingjang Emission recycling Creamer Reuters http://m.engineeringnews.co.za/article/startups-strive-to-recycle-emissions-for-new-carbon-economy-2019-05-31/rep_id:4433
23. Shell class action https://www.foei.org/press_releases/climate-legal-summons-submitted-shell
24. BBC .Honda confirms Swindon car plant closure' 19 February 2019 https://www.bbc.com/news/business-47287386
25. Documentary film 'American Factory' IMDB 2019 https://www.imdb.com/title/tt9351980/
26. Transparency International G20 https://www.transparency.org/news/feature/four_ways_the_g20_can_take_the_lead_on_anti_corruption
27. World Economic Forum https://www.weforum.org/agenda/2019/01/shaping-the-future-at-davos-2019/
28. Podcast sheree?? Climate Works Foundation https://www.climateworks.org/
29. Hydrocarbon Processing Worlds largest reverse osmosis plant https://www.hydrocarbonprocessing.com/news/2019/09/world-s-largest-reverse-osmosis-desalination-project and UAE EWEC https://wam.ae/en/details/1395302797149
30. Petroleum Economist CCGTs https://www.petroleum-economist.com/articles/midstream-downstream/power-

31. Hydrocarbon Processing Shell green hydrogen https://www.hydrocarbonprocessing.com/news/2020/02/shell-and-gasunie-plan-to-build-massive-dutch-green-hydrogen-plant
32. Cape Town water shortage https://www.weforum.org/agenda/2019/08/cape-town-was-90-days-away-from-running-out-of-water-heres-how-it-averted-the-crisis/
33. Corruption Perception Index 2018 https://www.transparency.org/en/cpi/2018
34. Hydrocarbon Processing Neste using wind power https://www.hydrocarbonprocessing.com/news/2019/10/neste-to-start-using-wind-power-at-its-production-sites-in-finland?id=3872815
35. Carbon trust https://www.carbontrust.com/resources/what-are-scope-3-emissions
36. Petroleum Economist Pavilion Singapore http://admin.petroleum-economist.com/articles/low-carbon-energy/energy-transition/2020/green-drivers-lead-to-innovation-for-pavilion
37. Petroleum Economist Net zero trend Anna Kachkova 21 February 2020 https://www.petroleum-economist.com/articles/low-carbon-energy/energy-transition/2020/major-trend-emerges-for-net-zero-targets
38. Solid Oxide Electrolysis Cell (SOEC) https://www.energy.dtu.dk/english/Research/Electrolysis-Cells/Solid-Oxide-Electrolysis-Cells
39. Yunus Social Business https://www.yunussb.com/
40. VW recycling https://www.volkswagenag.com/en/sustainability/environment/recycling.html
41. King IV report https://www.iodsa.co.za/page/DownloadKingIVapp
42. Housing https://www.geekwire.com/2019/microsoft-approaching-seattle-regions-affordable-housing-crisis/
43. Community upliftment Goedgedacht trust https://www.goedgedacht.org/
44. Mobile banking https://www.bbc.com/worklife/article/20131217-east-africa-a-mobile-banking-hub
45. Coral reefs https://sloactive.com/coral-reef-crisis-guide/
46. Saving turtles: Lang Tengah Turtle Watch http://www.langtengahturtlewatch.org/

47. Greenfins https://www.greenfins.net/about-green-fins
48. Oceana Why health oceans need sea turtles https://oceana.org/sites/default/files/reports/Why_Healthy_Oceans_Need_Sea_Turtles.pdf
49. McKinsey Plastics https://www.mckinsey.com/industries/chemicals/our-insights/how-plastics-waste-recycling-could-transform-the-chemical-industry#
50. Reforestation Willie Smits tedtalks How to restore a rainforest https://www.ted.com/talks/willie_smits_how_to_restore_a_rainforest?language=en
51. Tedtalk Earth for life Tobgay T https://www.ted.com/talks/tshering_tobgay_this_country_isn_t_just_carbon_neutral_it_s_carbon_negative?language=en
52. Neutrality: Oil and Gas: BP https://www.straitstimes.com/world/europe/bps-new-ceo-bernard-looney-goes-all-in-on-climate-goals-and-explores-overhaul
53. Atlantic Council May 21, 2020 Carbon capture and the Allam Cycle: The future of electricity or a carbon pipe(line) dream? by David Yellen https://www.atlanticcouncil.org/blogs/energysource/carbon-capture-and-the-allam-cycle-the-future-of-electricity-or-a-carbon-pipeline-dream/
54. Hydrocarbon Processing Expert Consortium unites for green Power-to-Fuel project to convert carbon emissions into cost-effective carbon neutral fuel. 3/25/2020 https://www.hydrocarbonprocessing.com/news/2020/03/expert-consortium-unites-for-green-power-to-fuel-project-to-convert-carbon-emissions-into-cost-effective-carbon-neutral-fuel?id=3872815
55. Hydrogen production breakthrough could herald cheap green energy Glasgow University https://phys.org/news/2014-09-hydrogen-production-breakthrough-herald-cheap.html
56. lhyfe https://lhyfe.com/en/hydrogene-lhyfe/
57. Petroleum Economist green hydrogen https://www.petroleum-economist.com/articles/low-carbon-energy/renewables/2020/green-hydrogen-can-be-cost-competitive
58. Nikola https://nikolamotor.com/motor
59. X prize https://carbon.xprize.org/prizes/carbon

60. World Bank Malaysia economy https://www.worldbank.org/en/country/malaysia/overview
61. China report China https://www.hrw.org/world-report/2019/keynote/autocrats-face-rising-resistance
62. China South Africa coal deal Limpopo coal concern https://www.thesouthafrican.com/news/china-south-africa-limpopo-coal-concern/
63. Frankopan P The New Silk Roads: The Present and Future of the World Bloomsbury 2018 https://www.amazon.com/New-Silk-Roads-Present-Future/dp/1526607425
64. Social progress Index www.socialprogress.org
65. Garratt B 'Stop The Rot' Routledge 2017 https://www.amazon.com/Stop-Rot-Reframing-Governance-Politicians/dp/1783537663
66. Net Zero Teeside Project https://www.netzeroteesside.co.uk/
67. Petroleum Economist climate change https://www.petroleum-economist.com/articles/low-carbon-energy/energy-transition/2020/climate-change-litigation-is-driving-the-transition
68. DOFE ideas https://www.dofe.org/do/ideas/
69. Protests Japan coal http://www.nocoaljapan.org/g20-protests-japan-end-coal/
70. BBC Global News Podcast Dr Ric Hoefnagels BBC World Service 25 October 2019. and Utrecht University Copernicus Institute of sustainable development https://www.uu.nl/en/research/copernicus-institute-of-sustainable-development
71. Energy policy US https://www.energypolicy.solutions/guide/
72. IEA https://www.iea.org/newsroom/events/publication-world-energy-outlook-2019.html
73. 4oceans https://4ocean.com/
74. African entrepreneur collective https://africanentrepreneurcollective.org
75. An Alternative Approach to Project Selection: The Infrastructure Prioritization Framework Darwin Marcelo, Cledan Mandri-Perrott, Schuyler House, Jordan Z. Schwartz World Bank PPP Group 14 April 2016 http://pubdocs.worldbank.org/en/844631461874662700/16-04-23-Infrastructure-Prioritization-Framework-Final-Version.pdf
76. WEF 2019 https://www.weforum.org/agenda/2019/01/shaping-the-future-at-davos-2019/

77. Contract for the web https://webfoundation.org/2019/03/web-birthday-30/
78. Global CCS Institute Global status of CCS 2019 report https://www.globalccsinstitute.com/resources/global-status-report/
79. Kotter J and Rathgeber H Our Iceberg is Melting: Changing and Succeeding Under Any Conditions Penguinrandomhouse 2006 https://www.amazon.com/Our-Iceberg-Melting-Succeeding-Conditions/dp/0399563911
80. Peoples report card https://www.youtube.com/watch?v=o08ykAqLOxk
81. Ref Bishop M, Green M. Philanthrocapitalism 2009 https://www.amazon.com/gp/product/1596916958/?tag=tedspeakers-20.
82. US government grants for renewable energy https://www.insidesources.com/us-still-subsidizing-renewable-energy-to-the-tune-of-nearly-7-billion/
83. Planet of the Humans Documentary film www.topdocumentaryfilms.com/planet-humans/
84. Net Zero Teesside Project https://oilandgasclimateinitiative.com/bp-eni-equinor-shell-and-total-form-consortium-to-develop-the-net-zero-teesside-project-and-accelerate-potential-of-uks-first-zero-carbon-cluster/
85. Project Management Institute (PMI) Agile Practice Guide https://www.pmi.org/pmbok-guide-standards/practice-guides/agile
86. Shell post covert change management https://www.hydrocarbonprocessing.com/news/2020/09/shell-launches-major-cost-cutting-drive-to-prepare-for-energy-transition?id=3872815
87. Amazon rainforest destroyed for sugar cane. https://news.mongabay.com/2019/11/sugarcane-threatens-amazon-forest-and-world-climate-brazilian-ethanol-is-not-clean-commentary/
88. Hyundai trucks for Switzerland https://www.hydrocarbonprocessing.com/news/2020/10/hyundai-delivers-first-fuel-cell-trucks-to-switzerland?id=3872815
89. Kahn Academy https://www.khanacademy.org/

90. Tedtalks countdown https://countdown.ted.com/
91. Danish energy policy https://www.irena.org/publications/2013/Jan/30-Years-of-Policies-for-Wind-Energy-Lessons-from-12-Wind-Energy-Markets
92. Philippi Economic Development Initiative (PEDI) http://pedi.org.za/
93. Carvajal A Elliott J 'Strengths and Weaknesses in Securities Market Regulation: A Global Analysis' IMF Working Paper WP/07/259 2007 https://www.imf.org/external/pubs/ft/wp/2007/wp07259.pdf
94. Angkor Wat https://www.nationalgeographic.com/news/2017/04/angkor-wat-civilization-collapsed-floods-drought-climate-change/#close
95. Climate Change Wisdom http://www.climate-change-wisdom.com/biosequestration.html
96. Wiki biosequestration https://en.wikipedia.org/wiki/Biosequestration#:~:text=Biosequestration%20is%20the%20capture%20and,continual%20or%20enhanced%20biological%20processes.
97. Kiss the Ground documentary film 2020 https://kissthegroundmovie.com/
98. Mongabay https://india.mongabay.com/2020/05/an-online-network-emerges-during-the-lockdown-connecting-farmers-directly-with-customers/
99. Harvesting Farmers Network (HFM) https://harvestingfn.com/
100. Sao Paulo water shortage https://50liters.com/sao-paulo-day-zero-lessons/#:~:text=In%202015%2C%20S%C3%A3o%20Paulo%20had,million%20people%20call%20it%20home.
101. Hurricane Katrina response https://en.wikipedia.org/wiki/Criticism_of_government_response_to_Hurricane_Katrina
102. US pandemic response crisis https://edition.cnn.com/2020/11/30/politics/federal-government-coronavirus-response-report/index.html
103. Obama pandemic preparedness systems https://www.vanityfair.com/news/2020/05/trump-obama-coronavirus-pandemic-response

## Chapter 7: The Future - What is Likely to Happen

1. PWC The World in 2050 https://www.pwc.com/gx/en/issues/economy/the-world-in-2050.html
2. Awakening the giants China and India https://podcasts.ox.ac.uk/assessing-economic-rise-china-and-india
3. Top ten largest economies https://www.valuewalk.com/2019/01/top-10-largest-economies-2050-china/
4. Bardhan P 'Awakening Giants, Feet of Clay: Assessing the Economic Rise of China and India' 2010 https://www.amazon.com/Awakening-Giants-Feet-Clay-Assessing/dp/0691156409
5. FT/IFC Transformational Business awards https://live.ft.com/Events/2020/FT-IFC-Transformational-Business-Awards-2020
6. Kroll bribery and corruption benchmarking https://www.kroll.com/en/insights/publications/compliance-risk/anti-bribery-and-corruption-benchmarking-report-2020
7. Dark web wikipedia https://en.wikipedia.org/wiki/Dark_web
8. Documentary Film 'Citizen Four' IMDB 2014 https://www.imdb.com/title/tt4044364/
9. Center for Disease Control (CDC) Atlanta https://www.cdc.gov/
10. Web 30th birthday https://webfoundation.org/2019/03/web-birthday-30/
11. Krypto currencies https://www.investopedia.com/terms/c/cryptocurrency.asp
12. US withdrawal from Northern Syria -Turkey invades. https://www.nytimes.com/2018/12/19/us/politics/trump-syria-withdrawal-obama.html
13. Johan Rockstrom Tedglobal 'Let the Environment Guide our Development Tedglobal 2010 https://www.ted.com/talks/johan_rockstrom_let_the_environment_guide_our_development?language=en
14. National Geographic Global Warming https://www.nationalgeographic.com/environment/global-warming/global-warming-overview/
15. Methane https://globalmethane.org/documents/gmi-mitigation-factsheet.pdf k

16. Hydrocarbon Processing CO2 Startups strive to recycle emissions for 'new carbon economy' 5/30/2019 https://www.hydrocarbonprocessing.com/news/2019/05/startups-strive-to-recycle-emissions-for-new-carbon-economy
17. SkyNRG https://skynrg.com/
18. Solar Hydrogen https://spectrum.ieee.org/energywise/energy/renewables/solar-closing-in-on-practical-hydrogen-production
19. UN Climate Progress Tracker tool https://unclimatesummit.org/trackerhome__trashed/trackerbusiness/
20. Neste petrochems https://www.neste.com/companies/products
21. Impossible foods https://impossiblefoods.com/
22. Biotech https://www.conserve-energy-future.com/biotechnology-types-examples-applications.php
23. Artificial Intelligence AI https://builtin.com/artificial-intelligence
24. Halung Bay Vietnam https://english.vov.vn/travel/ecofriendly-aquaculture-model-on-ha-long-bay-proves-fruitful-352208.vov
25. Aerosols https://earthobservatory.nasa.gov/features/Aerosols
26. Bloomberg Bowery and Bowery https://boweryfarming.com/
27. How to build Aquaponics https://learn.eartheasy.com/articles/how-to-grow-with-aquaponics-in-5-simple-steps/
28. Gladwell M The Tipping Point : How Little Things Can Make a Big Difference Little Brown and Company 2000 https://www.thoughtco.com/malcolm-gladwell-tipping-point-theory-3026765
29. World Bank The road to 2050 http://documents1.worldbank.org/curated/en/192421468341095824/pdf/360210rev0The0Road0to0205001PUBLIC1.pdf
30. Thorium https://whatisnuclear.com/thorium.html
31. Climate Action Tracker (CAT) https://climateactiontracker.org/
32. HSBC: The World in 2050: Quantifying the shift in the Global Economy https://warwick.ac.uk/fac/soc/pais/research/researchcentres/csgr/green/foresight/economy/2011_hsbc_the_world_in_2050_-_quantifying_the_shift_in_the_global_economy.pdf
33. Recycling plastics future Mckinsey

https://www.mckinsey.com/industries/chemicals/our-insights/how-plastics-waste-recycling-could-transform-the-chemical-industry#
34. UN Paris Accord Tracker and Wikipedia https://en.wikipedia.org/wiki/Carbon_neutrality
35. Frankopan P. The New Silk Road Bloomsbury 2019 https://www.amazon.com/New-Silk-Roads-Present-Future/dp/1526607425
36. Huawei https://coloradosun.com/2019/11/04/removing-banned-tech-from-chinas-huawei-will-cost-rural-colorado-telecoms-over-300-million-will-it-even-fix-the-problem/
37. Wellness Economy https://wellbeingeconomy.org/
38. BP Statistical Review of World Energy 2019 https://www.bp.com/content/dam/bp/business-sites/en/global/corporate/pdfs/energy-economics/statistical-review/bp-stats-review-2019-full-report.pdf
39. Woolworths carbon footprint https://www.woolworthsholdings.co.za/wp-content/uploads/2019/09/2018-Carbon-Footprint-Overview-.pdf
40. Cape Town water crisis Utube WEF https://www.weforum.org/agenda/2019/08/cape-town-was-90-days-away-from-running-out-of-water-heres-how-it-averted-the-crisis/
41. UAE EWEC https://wam.ae/en/details/1395302797149
42. Mongabay Saving Aru https://news.mongabay.com/2019/10/saving-aru-the-epic-battle-to-save-the-islands-that-inspired-the-theory-of-evolution/
43. Willie Smits https://www.ted.com/talks/willie_smits_how_to_restore_a_rainforest?language=en
44. PACE https://pacecircular.org/projects
45. World Economic Forum (WEF) 2019 https://www.weforum.org/agenda/2019/01/shaping-the-future-at-davos-2019/
46. DNV model https://eto.dnvgl.com/2020/highlights/?utm_campaign=GR_GLOB_20Q3_PROM_ETO_2020_Event_FAQ_Article_Others_B&utm_medium=email&utm_source=Eloqua#forewordlaunch
47. UN climate change tracker https://docs.google.com/spreadsheets/d/1uBjjcIsB2ommkarTabFhHa7Nrl

KSRhLf1HHeGe9M6PQ/edit#gid=865795372
48. Wellbeing Government Alliance https://wellbeingeconomy.org/wego .
49. NASA ozone hole https://www.nasa.gov/feature/goddard/2019/2019-ozone-hole-is-the-smallest-on-record-since-its-discovery Blank Ibid copy of no. 45
50. Malawi: Bureau of Investigative Journalism https://www.thebureauinvestigates.com/stories/2018-08-08/scourge-superbugs-killing-babies-malawi
51. CNBC 'Why Facebook's Libra Cryptocurrency is in trouble' https://www.youtube.com/watch?v=vPu4kn5GN5M
52. World Bank Windhoek water https://blogs.worldbank.org/water/in-windhoek-iuwm-is-key-to-closing-the-water-loop
53. IEAs Energy Technology Perspective (ETP) https://www.iea.org/reports/energy-technology-perspectives-2020
54. C40 Cities https://www.c40.org/about
55. Vanadium redox battery https://en.wikipedia.org/wiki/Vanadium_redox_battery
56. BBC podcast Perovskite, the new material experts say will transform solar power https://www.bbc.co.uk/sounds/play/p02qm977
57. UN Paris Accord Tracker/Companies/Cement
58. Marine biofuel Hydrocarbon Processing https://www.hydrocarbonprocessing.com/news/2019/10/shipping-companies-retailers-look-to-develop-cleaner-marine-biofuel?id=3872815
59. Sailing ships https://theweek.com/articles/825647/why-cargo-ships-might-literally-sail-high-seas-again
60. Fuller GE A world without Islam Back Bay Books Little Brown Company 2010 https://www.amazon.com/World-Without-Islam-Graham-Fuller/dp/031604119X
61. HSBC https://enterprise.press/wp-content/uploads/2018/10/HSBC-The-World-in-2030-Report.pdf
62. IEA IGCC https://www.iea-coal.org/integrated-gasification-combined-cycle-igcc/
63. Pretoria Portland Cement (PPC) https://www.ppc.africa/za/strength-beyond/going-beyond/tyre-burning-initiative

64. Lanzatech https://www.lanzatech.com/2018/06/08/worlds-first-commercial-waste-gas-ethanol-plant-starts/
65. Geneco https://www.geneco.uk.com/Liquidwaste/
66. Nano technology https://phys.org/news/2016-03-ways-nanotechnology-future.html
67. IEA world energy outlook 2020 https://www.iea.org/reports/world-energy-outlook-2020/overview-and-key-findings
68. IEA Energy Technology perspectives 2010 https://www.iea.org/reports/energy-technology-perspectives-2010
69. Rosling H Factfulness: Ten Reasons We're Wrong About the World--and Why Things Are Better Than You Think Hodder and Stoughton 2018 https://www.amazon.com/Factfulness-Reasons-World-Things-Better/dp/1250107814 .
70. Millenium Project http://www.millennium-project.org/projects/challenges/
71. Millenium project challenges https://www.youtube.com/watch?v=J5fof-Qt4nI&list=PL_LEFZFKTl5HkzUe_vOYXkwjK7-Z0ix6N&index=7
72. OECD Building Back Better: a sustainable resilient recovery after Civid-19 https://read.oecd-ilibrary.org/view/?ref=133_133639-s08q2ridhf&title=Building-back-better-_A-sustainable-resilient-recovery-after-Covid-19
73. Wellbeing Economies Film 2020 https://lorenzofioramonti.org/videos/
74. Crowther lab https://www.crowtherlab.com/restoration/
75. Tedtalks Thomas Crowther biodiversity restoration https://www.ted.com/talks/thomas_crowther_the_global_movement_to_restore_nature_s_biodiversity
76. Electricity sustainability Bent Erik Bakken 13 January 2020 Petroleum economist https://www.petroleum-economist.com/articles/midstream-downstream/power-generation/2020/electricity-production-is-on-a-sustained-charge
77. International Civil Aviation Organization (ICAO) 26 October 2017 http://sdg.iisd.org/news/icao-conference-agrees-on-2050-vision-for-sustainable-aviation-fuels/
78. Report Ammonfuel: An industrial view of ammonia as a marine fuel Alfa Laval, Hafnia, Haldor Topsoe, Vestas, Siemens Gamesa August 2020

https://info.topsoe.com/ammonfuel

79. Prescott-Allen R The Wellbeing of Nations- A Country-by-Country Index of Quality of Life and the Environment Island Press 2000 https://www.amazon.com/Wellbeing-Nations-Country-Country-Environment/dp/1559638303
80. The Ocean Agency https://theoceanagency.org/
81. Documentary film Chasing Coral ' Orlowitz J Netflix 2016 https://www.netflix.com/za/title/80168188
82. Corona virus Thai Enquirer https://www.thaienquirer.com/14192/opinion-we-cannot-return-to-business-as-usual-after-the-pandemic/
83. IEA post Corona recovery plan https://www.iea.org/news/chair-s-summary-for-iea-clean-energy-transitions-summit
84. IEA quote https://www.petroleum-economist.com/articles/low-carbon-energy/energy-transition/2020/emissions-past-peak-in-ieaimf-green-recovery-plan
85. Copenhagen sustainable energy https://www.hydrocarbonprocessing.com/news/2020/08/haldor-topsoe-joins-ambitious-sustainable-fuel-project-in-denmark
86. The World Counts https://www.theworldcounts.com/stories/Chemical_Pollution_Examples \
87. North Dacota State University fertilzers https://www.ag.ndsu.edu/publications/environment-natural-resources/environmental-implications-of-excess-fertilizer-and-manure-on-water-quality
88. David Attenborough: A Life on Our Planet Netflix Documentary Film 2020 https://www.netflix.com/title/80216393
89. State of Surveillance documentary film 2016 https://topdocumentaryfilms.com/state-surveillance/
90. 'The Social dilemma' documentary drama hybrid film 2020 Netflix https://www.netflix.com/za/title/81254224
91. 'The Great Hack' documentary film 2019 Netflix https://www.netflix.com/za/title/80117542
92. IEA Energy Technology Perspective (ETP) launch

https://www.iea.org/news/reaching-energy-and-climate-goals-demands-a-dramatic-scaling-up-of-clean-energy-technologies-starting-now

93. Wellbeing economy https://wellbeingeconomy.org/
94. GDP announcement The Guardian https://www.theguardian.com/business/2020/aug/12/uk-economy-covid-19-plunges-into-deepest-slump-in-history
95. Ref Raworth K 'Doughnut Economics: Seven Ways to Think Like a 21st-Century Economist' 2017 https://www.amazon.com/Doughnut-Economics-Seven-21st-Century-Economist/dp/1603586741
96. Kate Rayworth website https://www.kateraworth.com/doughnut/
97. Restor biodiversity platform https://www.crowtherlab.com/restoration/
98. Documentary film 'The brave blue world: Racing to solve our water crisis' IMDB 2019 https://www.imdb.com/title/tt11921004/
99. Why Chennai has run out of water BBC https://www.bbc.com/news/world-asia-india-48797399
100. Paypal crypto https://edition.cnn.com/2020/10/21/investing/paypal-bitcoin-cryptocurrencies/index.html
101. WEF The Future of Jobs Report 2020 https://www.weforum.org/reports/the-future-of-jobs-report-2020
102. Kiss the Ground documentary film 2020 Netflix https://www.netflix.com/title/81321999
103. Hydrogen Fuel Cell Trains Hydrocarbon Processing https://www.hydrocarbonprocessing.com/news/2020/11/siemens-deutsche-bahn-launch-local-hydrogen-trains-trial?id=3872815
104. Extreme weather https://e360.yale.edu/digest/extreme-weather-events-have-increased-significantly-in-the-last-20-years
105. EPA study on fracking and drinking water. https://www.epa.gov/hfstudy
106. China crypto currency https://www.bloomberg.com/news/articles/2019-08-12/china-s-pboc-says-its-own-cryptocurrency-is-close-to-release
107. On line shopping stats https://www.bigcommerce.com/blog/online-shopping-statistics/
108. Tedtalk Kate Raworth https://www.youtube.com/watch?v=Rhcrbcg8HBw&ab_channel=TED
109. Varshini Prakash Sunrise Movement https://podcasts.apple.com/za/podcast/how-i-built-this-with-guy-

raz/id1150510297?i=1000497277238

# ABOUT THE AUTHOR

Robert Bruce

Robert Bruce, the fighter, fought for the independence of Scotland from England, and subsequently became King of Scotland.

Robert Bruce, the writer, is fighting for a fairer world where merit prevails over corruption and greed. He is the author of two books and spends his reclusive days writing and undertaking environmentally friendly projects.

The author can be contacted at deathofbusinessasusual@gmail.com

Made in the USA
Middletown, DE
27 February 2022